EXPLORING WITH A MAGNIFYING GLASS

EXPLORING WITH A MAGNIFYING GLASS

by Kenneth G. Rainis

Franklin Watts
New York/London/Toronto/Sydney
1991
A Venture Book

On the cover: *Volvox globator,* magnified 10x.
Copyright © Bruce Russell.

Diagrams by: Vantage Art (pp. 13, 14, 15, 18, 20, 23,
25, 27, 36, 42, 77, 110, 128) and Kenneth G. Rainis (pp. 46, 58,
60, 62, 66, 69, 72, 78, 80, 82, 88, 89)

All photographs copyright © Bruce J. Russell, BioMEDIA Associates

Library of Congress Cataloging-in-Publication Data

Rainis, Kenneth G.
Exploring with a magnifying glass / Kenneth G. Rainis.
p.   cm. — (A Venture book)
Includes index.
Summary: Presents activities using a magnifying glass and, for the
most part, objects that are found around the house.
ISBN 0-531-12508-4
1. Magnifying glasses—Experiments—Juvenile literature.
2. Science—Experiments—Juvenile literature. [1. Magnifying
glasses—Experiments. 2. Science—Experiments. 3. Experiments.]
I. Title.
QC373.M33R35   1991
507.8—dc20   91-18329 CIP AC

CONTENTS

ACKNOWLEDGMENTS

This book could not have become a reality without the support and encouragement of my wife, Joan, and my children, Michael and Caroline.

Tom Cohn, my editor, gave me the freedom to explore and develop the ideas that are presented here.

Special mention of the individuals who helped give this book an added dimension: Ernie Jeneault and David Hackworth of Standard Register for their support and cooperation in making available the special and unique printed images. My colleague Howard Kimel, of Kemtec Educational Corp., who provided support in the forensic sciences. Steve Bryson who gave quick lessons in geology and paleontology and helped make these sections understandable for young people. Stephanie Miller and Gervase Pevarnik, ever supportive, assisted in creating the special magnification and three-dimensional imaging photographs. Robert Iveson, my friend and colleague, reviewed the manuscript with a critical eye always toward "making sure it's fun"!

Special thanks goes to Bruce Russell of BioMEDIA Associates, whose many photographs, including the cover, will help young readers visualize just what exploring with a magnifying glass is truly like!

In memory of our son,
Billy Rainis

1

THE MAGNIFYING GLASS

The *magnifying glass* is a tool that allows you to better understand the natural world around you. Its only limitation in exploring this world is the limits of your curiosity.

This book contains activities that center around the major components of our natural world—images, surfaces, minerals, and living things, including yourself. Each activity is not an end in itself, but a beginning, an opportunity to explore and to learn.

These activities require only the interest of the moment that I hope will lead to a more thorough investigation later on. In many instances I have expanded activities into long-term projects that can serve as the basis for a science fair project. Some activities are designed for you and a friend to share together. My hope is that you will have as much fun in doing the activities as I did in sharing them with you!

Here are just some of the questions you can answer by using a "looking" glass:

• How is thread made?

- What makes up dust?
- What does moldy bread look like?
- Is the outer edge of your fingernail layered?

Almost all of the materials needed for these activities can be found in and around your home. It is a good idea to have a notebook close at hand to jot down notes about your observations. Sources for a few unique materials are listed in the appendix in the back of this book. The essential tool, of course, is a magnifying glass . . . and your curiosity!

SAFETY—THE MOST IMPORTANT PART IN ANY EXPLORATION

Eleven rules to keep safe:

1. Be serious about science. A glib attitude can be dangerous to you and to others. Be responsible; do not use a magnifying glass to burn materials or to start fires.
2. **Never** look directly into a lens pointed directly at the sun. Doing so can cause serious injury to your eyes.
3. Read instructions carefully before starting any activity outlined in this book. If you plan to expand on these activities, DISCUSS YOUR EXPERIMENTAL PROCEDURE WITH A SCIENCE TEACHER OR A KNOWLEDGEABLE ADULT before going ahead.
4. Keep your work area clean and organized. Never drink or eat while conducting experiments.
5. Respect all life-forms. Never mishandle any living vertebrate. Never perform any experiment on a live vertebrate that will injure or harm the animal.
6. Wear protective goggles when doing activities in-

volving chemicals, when heating objects or water, or when performing any other experiment that could lead to eye injury.

7. Do not touch chemicals with your bare hands unless instructed to do so. Do not taste chemicals or chemical solutions. Avoid inhaling vapors or fumes from any chemical or chemical solution.

8. Clean up any chemical spill immediately. If you spill anything on your skin or clothing, rinse it off immediately with plenty of water. Then report what happened to a responsible adult.

9. Keep flammable liquids away from heat sources.

10. **Never** operate a power tool without direct adult supervision.

11. **Always** wash your hands after conducting activities. Dispose of contaminated waste or articles properly.

MAKING A "LOOKING" GLASS

A magnifying glass is a lens which makes close objects appear larger. For centuries scientists have made lenses from transparent materials having at least one curved surface, and have used those lenses to form magnified or reduced images or to concentrate or spread light rays. Probably the earliest published use of a glass magnifier was by the English philosopher and scientist Roger Bacon (1214–1294), who used a spherical glass magnifier for reading. You can duplicate his experiment by using a clear glass marble. But first, why not make your own magnifying glass?

Take a drinking glass and fill it almost to the top with water. Now take this book and place it behind your homemade "looking" glass. Look through the center area of the glass. Are the printed words on this page magnified?

Most lenses are made of glass, but the one you just

made was a water lens. Let's make another water lens. Wrap and twist a piece of thin copper wire around the shank of a large nail to form a loop. See Figure 1. Dip this loop into water and look through it. Place the lens close to your eye for viewing. Look at printed words. Are individual letters enlarged? Are they right side up or inverted? Although this type of lens is delicate (the water film can easily break), use it to view other familiar objects: flowers, small insects, printed photographic images, etc. Compare the magnifying ability of your water lens to that of other, glass lenses. Are there differences? (See the next two sections of this chapter.)

HOW A MAGNIFYING GLASS WORKS

Pick up a typical magnifying glass and examine it. Most likely, both surfaces of the lens are convex, or curve outward. (If the surface of a lens curves inward it is termed *concave.)* The *double-convex* magnifying lens you are probably holding gives two kinds of images. Hold the magnifying glass close to this page. The upright, enlarged image you see is called a *virtual image.* The light rays which produce this image diverge (spread out) as they pass through the lens and appear to originate on the same side of the lens as this printed page. A virtual image is always right side up. Use Figure 2A as a guide in understanding how a double-convex lens creates a virtual image.

Now move the magnifying glass further away from this page until the printed words appear upside down. This *real image* is formed when light rays from an object pass through the lens and are focused on the other side of the lens. A real image is always upside down. Use Figure 2B as a guide in understanding how a double-convex lens creates a real image.

Is the image you see in a mirror a virtual or real image?

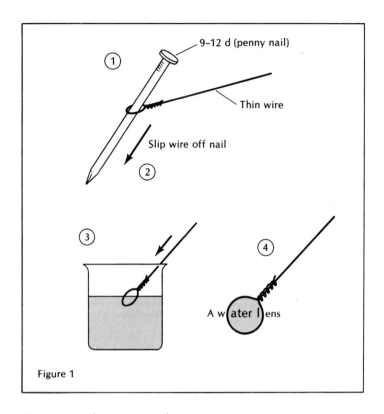

9-12 d (penny nail)

①

Thin wire

Slip wire off nail

②

③

④

A water lens

Figure 1

How to make a water lens

The magnifying power of a lens depends upon its *focal length* —the distance from the center of the lens to the point where the image is focused on a surface (such as a white card). A lens with greater curvature has a shorter focal length and thus a greater magnifying ability. It refracts (bends) light rays more toward each other, and thus they meet at a smaller distance from the lens. Use Figure 2C as a guide to how a double-convex lens reflects light rays.

13

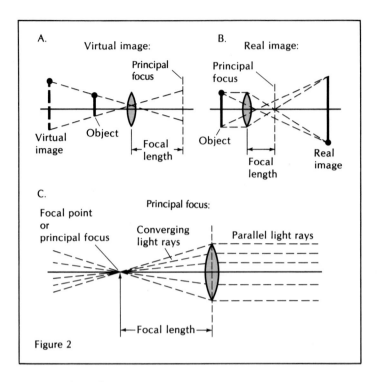

How a lens images

ESTIMATING THE MAGNIFYING ABILITY OF A LENS

Magnification is the ratio of the apparent size of a magnified object to its true size. The greater the magnifying ability of a lens, the shorter its focal length; the shorter the focal length, the greater the magnification.

Most people hold a book (or any other object) about 10 inches (25 cm) away when reading. This 10-inch distance is considered for this discussion to be the focal length of the eye. This ratio can be expressed as:

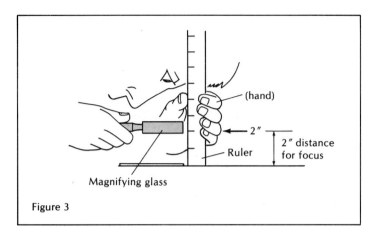

Figure 3

Calculating the focal length of a lens

$$\text{Magnification} = \frac{\text{focal length of human eye}}{\text{focal length of lens}} = \frac{10 \text{ inches}}{X \text{ inches}}$$

For example, suppose your hand lens has a focal length of 2 inches, or 5 cm (see Figure 3). Its magnification would be calculated this way:

$$\text{Magnification} = \frac{10 \text{ inches} \quad \text{(eye reading distance)}}{2 \text{ inches} \quad \text{(focal length of hand lens)}}$$
$$= \quad 5 \text{ X}$$

A hand lens having a focal length of 4/8 inch (or 0.5 inch) would have a magnification of 20X. Suppose you had a lens with a focal length of 3/8 inch; what would be its magnification?

A GUIDE TO "LOOKING" GLASSES

Whatever the name—hand lens, pocket lens, magnifying glass—lenses come in a variety of designs and magnifying powers (see photo on p. 17). Listed below is a quick-reference guide to the types available, their magnification ranges, and suggested uses for activities in this book.

- *Hand lens–reading glass.* Magnification range usually from 2.5X to 10X. Large diameter. Useful for wide-field viewing or in constructing optical devices.
- *Folding pocket magnifiers.* Usually from 4X to 20X. Pocket magnifiers can be a single lens or a combination of up to three lenses. Multiple-lens pocket magnifiers have lenses that can be used separately, or in combination for higher magnification. Useful for close-field viewing.
- *Hastings triplet magnifier* (expensive). Either 10X or 20X. Three individual lenses are fused together to form a single glass element giving a flat image without the fuzziness around the edges that single double-convex magnifying lenses have. Small-diameter lens for close-field viewing.
- *Jeweler's loop.* Usually 10X. Lens cup held by the eye for close-up viewing; hands are free for manipulating the object to be viewed.
- *Stand magnifier.* Either 5X or 10X. Placed over small object to be examined for close-field viewing.

These "looking" glasses can be obtained in photography and stationery stores or through the science-supply companies listed in the appendix.

LOOKING THROUGH A FLAT LENS—THE FRESNEL

A Fresnel lens (pronounced fray-nel) is a flat, thin piece of optically clear plastic; on one side are small molded concentric grooves (prisms) of various heights. These stepped zones extend from the center to the outer margins (Figure 4). Each groove is in fact a precisely molded isosceles

Magnifying lens types: (A) Triplet magnifier. (B) Jeweler's loop. (C) Double-lens tripod stand magnifier ready to examine some duckweed (Lemna). (D) Hand/reading lens. (E) Folding pocket magnifier. (F) Fresnel lens.

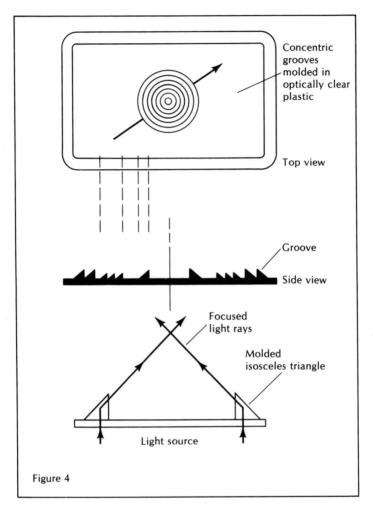

Concentric grooves molded in optically clear plastic

Top view

Groove

Side view

Focused light rays

Molded isosceles triangle

Light source

Figure 4

Examining a Fresnel lens and how it magnifies

triangle. Acting together, these specially molded re-fracting prisms bend light rays inward like a traditional double-convex glass lens.

Fresnel lenses are available at stationery stores and

through science-supply companies. Does the Fresnel lens have the same optical properties as a glass lens? For example, can it form virtual and real images? Try producing both types of images with a Fresnel lens.

Note: **Because they concentrate light rays, these lenses must be used with caution around direct sunlight!**

Use a traditional magnifying glass to observe and count the molded concentric rings in a Fresnel lens. If possible, compare the number of lines per inch among Fresnel lenses of different magnification. Is there a difference?

Besides being used as magnifiers, Fresnel lenses are image and light concentrators. Invented by Augustin Jean Fresnel (1788–1827), the lens was first used to concentrate light produced from oil lamps in lighthouses. Investigate an overhead projector at school. Is the Fresnel lens at work today?

FOR BETTER VIEWING: RESTRICT THE LIGHTING!

The great microscopist Anton van Leeuwenhoek (1632–1723) used to use light from a window (with a southern exposure) when viewing through his powerful single-lens microscope. Such a small light source, he said, gave a clearer image.

Here are some tips for better viewing:

- Always try to reduce the cone of light that illuminates the object you are examining. You can do this by using a window, standing under a tree, or using a single light source, such as a desk lamp, when viewing. See Figure 5A. Position yourself so that you are looking at right angles to the light source. I call this viewing technique *side lighting.*
- Sometimes *direct lighting* (a desk lamp near the object

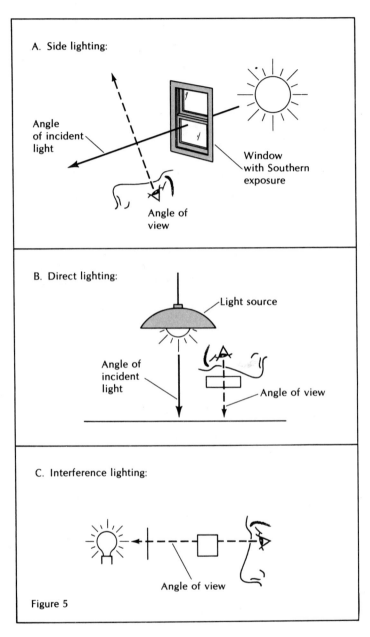

A. Side lighting:

Angle of incident light

Angle of view

Window with Southern exposure

B. Direct lighting:

Light source

Angle of incident light

Angle of view

C. Interference lighting:

Angle of view

Figure 5

Magnifying-glass viewing methods

to be viewed) is equally effective, especially when examining the printed word or a graphic image such as an engraving or photograph (Figure 5B). Such a "point source" provides a large amount of directed illumination.

- Another technique is to use *interference lighting* (Figure 5C). Here the object to be viewed is placed directly in front of a light source when viewed with a lens. For example, hold an eggshell fragment up to a strong light to observe minute pores that allow for oxygen diffusion. Photographic negatives or transparencies require this type of lighting.
- Light-colored objects are best viewed over dark backgrounds (or in dark shadows). Similarly, dark-colored objects require light backgrounds.
- Very small objects (dead insects and such) are best viewed by holding them with tweezers or impaled by a pin.
- Do not remain still when peering through your lens. Move around! Try various lighting conditions and situations. Hold the object to be viewed in one hand and your magnifying glass in another. Constantly vary the illumination levels to obtain the best balance of light contrast for viewing.

Always remember: *Never* look through any lens pointed directly at the sun. It can severely damage your eyes!

BUILDING MICROSCOPES

A microscope is an optical instrument consisting of a single lens or a combination of lenses for viewing enlarged images of objects. Microlife forms such as protists (protozoans and algae) and microinvertebrates (like

21

Daphnia and rotifers) are best viewed through single-lens microscopes. (A compound, or multiple-lens, microscope can also be used to view microlife forms. These instruments provide magnification ranges up to 1000X!)

In this section are two methods of building a single-lens microscope. One uses an ordinary 10X or 20X pocket magnifier; the other, a lens with much more magnifying power—a copy of the type used by Leeuwenhoek himself!

THE MAGNIFYING-GLASS MICROSCOPE

What You Need
> safety goggles (to protect your eyes)
> hand or electric drill with 3/4-, 1/4-, and 1/16-inch
>> bits
> 6 inches of 1-inch dowel
> 2 1/2 inches of 1/2-inch dowel
> 1 1/2 inches of 1/4-inch dowel
> white glue
> Velcro strip, 1/2 inch by 1 inch
> 2 rubber bands
> canvas needle, 2 inches long
> pocket magnifier (10X or 20X)

What to Do
> Use the instructions listed below and Figure 6 as a guide to building this microscope. Also, be sure that a responsible adult is there to supervise as you work.

1. **Put on your safety goggles.**
2. Drill a 3/4-inch hole through the 1-inch dowel about 2 inches from the end.
3. Glue a strip of Velcro to line this hole.
4. Use the rubber bands to attach the pocket magnifier to the drilled end of the 1-inch dowel.
5. Drill a 1/4-inch-diameter hole 1/2 inch from one end

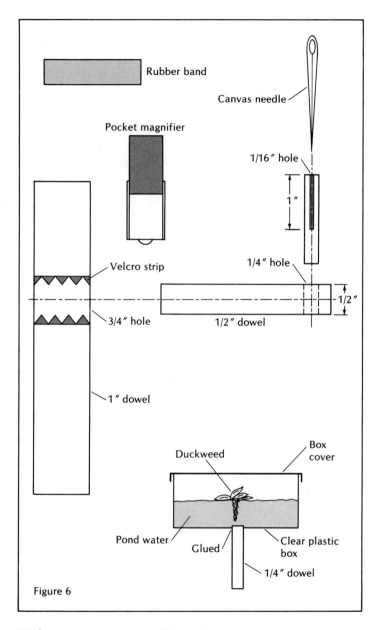

Rubber band

Canvas needle

Pocket magnifier

1/16″ hole

1″

1/4″ hole

1/2″

Velcro strip

3/4″ hole

1/2″ dowel

1″ dowel

Duckweed

Box cover

Pond water

Glued

Clear plastic box

1/4″ dowel

Figure 6

Make your own magnifying-glass microscope

of the 1/2-inch dowel. Drill the hole all the way through the dowel.

6. Drill a 1/16-inch hole in the face cut of one end of the 1/4-inch dowel and insert the canvas needle into this hole.

7. Insert the 1/4-inch dowel into the hole in the 1/2-inch dowel.

8. Insert the 1/2-inch dowel into the 1-inch dowel so that the assembly faces the side where the magnifier is attached.

Figure 7 is a drawing of the completed magnifier microscope.

Using Your Magnifying-glass Microscope

Pick up water samples by dipping the eye of the needle into a collected water sample—from a puddle, pond, aquarium, almost anywhere! Replace the needle in the hole of the 1/4-inch dowel and position it in line with the lens by manipulating the dowels. Use your thumb and index finger to bring the microscope into sharp focus by moving the 1/2-inch dowel.

Begin with direct lighting when using this microscope. Stand about 3 feet (1m) from a window, or experiment with distance and amount of light!

You can adapt the microscope to hold other objects (flowers, insects, etc.) by using the sharp point of the needle. Simply reverse the needle in the hole of the 1/4-inch dowel. To view large volumes of a water sample or microlife forms on or around aquatic plants, cement a clear, rectangular plastic box (with cover) to a 1/4-inch dowel as a replacement for the needle holder.

THE BEAD-LENS MICROSCOPE

This project is only for the more experienced young microscopist. It **must** be done under adult supervision.

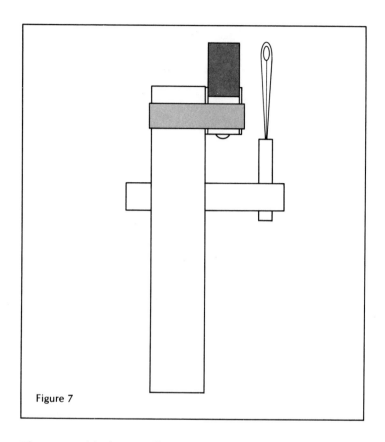

Figure 7

The assembled magnifier microscope

What You Need
    goggles (to protect your eyes)
    12 inches of flint glass rod (3-mm outer diameter)
    2 potholders
    hammer
    power or hand drill
    1/16-inch metal drill bit
    metal file
    2 pairs of pliers

handkerchief
rectangular piece of thin aluminum or sheet metal
  (3 × 6 inches) with rounded corners
modeling clay
canvas needle
1/4" wooden dowel, 3 inches long

What to Do

Use the instructions listed below and Figure 8 as a guide to constructing this microscope.

1. **Put on your safety goggles.**
2. Start by making a bead-glass lens. Use the potholders to grasp each end of the glass rod. Position the **center** of the rod about 1 inch over a kitchen-stove gas-burner flame. (An electric stove cannot be used.) Slowly rotate the rod. As it heats, it will start to bend. When this happens, draw a glass thread by **gently** moving the ends of the rod away from each other.
3. Carefully heat the longer of the two glass threads. Do so by bringing its edge over a **low** burner flame until the glass starts to ball up. Let the sphere grow until it is about 1/8-inch (3 mm) in diameter. Allow it to cool. After cooling, wrap the glass bead in a handkerchief. Use pliers to grasp both the sphere and the rod and gently snap the sphere off the rod. Carefully unwrap the glass bead from the handkerchief.
4. Carefully bend the 3 × 6-inch (75 × 152 mm) rectangular sheet metal in half, lengthwise.
5. Drill a 1/16-inch aperture hole about 1/2 inch from the bottom of the metal fold. *(Note:* Any aperture

Making a bead-lens microscope

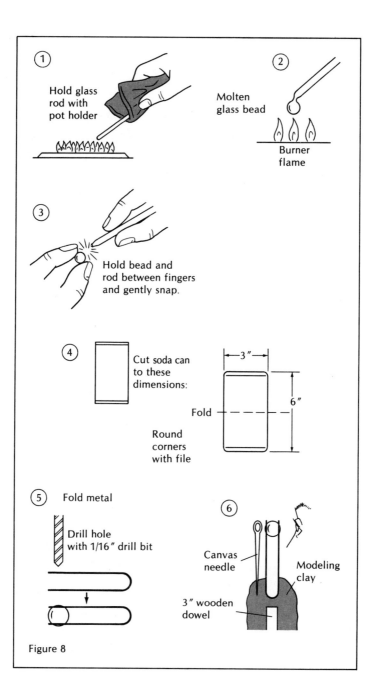

① Hold glass
rod with
pot holder

② Molten
glass bead

Burner
flame

③ Hold bead and
rod between fingers
and gently snap.

④ Cut soda can
to these
dimensions:

← 3″ →

6″

Fold

Round
corners
with file

⑤ Fold metal

Drill hole
with 1/16″ drill bit

⑥

Canvas
needle

Modeling
clay

3″ wooden
dowel

Figure 8

hole should be half the diameter of any bead lens.)

6. Use the metal file to smooth any rough edges from either drilling or metal cutting.

7. Carefully mount and set the lens between the two aperture holes made in the sheet metal, bending the open end over with pliers as necessary to hold the lens in place.

8. Mold modeling clay around both the 3-inch wooden dowel (for added support) and the metal fold to make a handle.

9. Dip the eye of the canvas needle into the water sample to be explored. Carefully insert the point of the needle into the clay *very close* to the lens (almost touching, about 1 mm, or *less);* the needle's eye must be in line with the center of the spherical lens.

Using Your Bead-Lens Microscope

A 3-mm bead lens has an extremely short focal length, producing magnification values of close to 100X! For illumination, begin with a "point source" of light such as a flashlight in a darkened room placed at eye level, about 3 feet (1 m) distant.

Hold the instrument *very close* to your pupil to observe the image. These conditions will produce a grainy image, but the improved contrast will let you see without difficulty. Once you are used to viewing through a bead lens, you can switch to other, more diffused lighting conditions such as a window (having southern light) for viewing minute detail.

This type of microscope was used by Leeuwenhoek in his myriad observations of microlife that he recorded in letters and drawings he sent to the Royal Society in London. Chapter 3 gives you a guided tour of this World of the Little.

READ MORE ABOUT IT

Berger, M. *Lights, Lenses, and Lasers.* New York: Putnam, 1987.

Bleifeld, M. *Experimenting with a Microscope.* New York: Watts, 1988.

Headstrom, R. *Adventures with a Hand Lens.* New York: Dover, 1976.

2
A CLOSER LOOK
AT IMAGES

An *image* is the imitation, likeness, or graphic representation of an object. Images are powerful communication mechanisms around which the whole cultural fabric of our world revolves. This chapter will provide you with a guide on how images are made—and how to make your own!

PRINTED IMAGES

BLACK-AND-WHITE PHOTOGRAPHS

Unlike line drawings, black-and-white photographic images cannot be printed unless the image itself is broken up into dots. Use your lens to study a printed black-and-white photograph. Notice that larger dots compose dark areas and very small dots form lighter areas. Study a variety of photographs found in calendars, newspapers, magazines, and in this book.

This dot effect is called *screening,* and the resultant

image is termed a *halftone*. In halftones a screen (a piece of clear film containing hundreds of closely spaced grid lines) is placed directly in front of the photographic film exposed when the original black-and-white photograph is copied. The grid lines break the picture into dots, producing a halftone negative used to produce the image itself.

Screens can be coarse or fine, depending upon the paper the image is printed on, as well as the quality of the desired image. Locate a black-and-white photograph in a magazine. Use a ruler to mark off either a 1-inch or a 1-centimeter segment of the photograph that has an even gray tone. Use a magnifying glass to count the number of rows of dots within the measured area. If you marked off a 1-inch segment, you should count 133 rows (or staggered dots). This is a 133 (per inch) screen or 54 (per centimeter) screen. Use the same technique to count the screen of a black-and-white photo in a newspaper. Is it different? Try determining the screen of a similar image in an art book (but do not mark up the book in any manner). Is it finer (more rows) that the average 133 screen? What screen do the photographs in this book have?

COLOR PHOTOGRAPHS

Color photographic images have to be 'separated'— that is, the effect of full color in print is achieved by breaking the original image into three component colors, plus black. Each of these colors is individually screened by photographing it through an appropriate color filter. These four screened color images are combined during printing to produce the color image you see. Printers call this process "four-color."

• Find a printed color photograph in a magazine, one that has a wide range of colors. Use the magnifying glass to observe multiple-color areas. Can you

31

identify the three primary colors (in addition to black) used in the printing process? What are they? Can you observe the positioning of the various screens (rows of individually colored dots) to create other colors?

- Locate color panels (called *tints*) in a magazine or book. (For example, look for colored borders or colored areas around blocks of type at the beginning of a magazine article.) Use the magnifying glass to find out whether it is a solid color or a screen. If screens are present, are they the same as the screen in a printed black-and-white photograph? Are tints composed of one or many colors?
- Examine various printed color photographs to determine if they were made by either a two-, three- or four-color process.
- Sometimes the printer makes an error and the resulting color photograph appears blurred. Use the magnifying glass to examine such an example.
- Use the Yellow Pages to see if there is a printer near where you live. Visit his/her establishment and ask for samples called "press proofs" of color printing. Use the magnifying glass to examine the multicolored bars at the bottom of a press page. Printers use these bars to adjust the amount of colored ink used in the printing process.

PHOTOGRAPHIC IMAGES

These permanent images are made on light-sensitive materials. This process involves using a glass lens to focus the image onto a light-sensitive film emulsion containing crystals of silver bromide suspended in a transparent gelatin. Crystals of silver bromide are transparent until they are exposed to light, which turns them black. Following exposure, the film is chemically processed (developed) both to preserve the silver (converted by the reaction of silver bromide with light) on the emulsion surface and to wash off any excess silver bromide. Highly reflective (light) areas of the image will convert more silver bromide to silver in the film emulsion, leaving a dark area on the film.

Conversely, nonreflective (dark) areas of the image will leave little or no silver. Thus the light image recorded on the film is a "negative," or reversed image of the subject photographed. When light (from a light bulb in an enlarger) passes through a negative and falls on an equally sensitized photographic paper, the result is a "positive," or normal, image—a photograph!

- Use the highest-power magnifying glass you have (at least 10X) to observe a black-and-white photograph under direct lighting conditions. Look to see the randomly distributed dark grains of silver that make parts of the image black or gray. Can you tell where there is almost no silver deposited?
- Hold a black-and-white photographic negative up to a light source. Again examine it using the highest power magnifying glass you have (at least 10X). Can you observe dark grains of silver?
- A film's ability to react to light and produce an image is termed *film speed.* High-speed films have more silver bromide crystals than films of slower speed and are thus able to react to smaller amounts of light. For example, you would use a high-speed film to photograph a car race, or to capture an image in a low-light situation.
- If possible, compare the overall amount of silver grains present in a negative of a high-speed film (e.g., Kodak TRI-X) to that of a low-speed film (Kodak Panatomic-X). Try the same comparison with color films—compare the film negative "grain" of a 1000 ASA high-speed film with a 60 ASA film.

BUILDING THE CAMERA OBSCURA,
THE PENCIL OF NATURE

Darken all the windows in a room except one. In one hand hold a magnifying lens so that one of its convex surfaces is oriented toward the scene outside. Use your other hand to hold a white piece of paper against the opposite convex lens surface. Slowly move the paper away from the lens. At some distance you will be able to observe the image of the outdoor scene projected onto the paper. How is the image oriented? Note the distance from the lens to the paper (the lens's focal length) at a point where you observe its sharpest image.

In the 1800s artists used an optical device called the *camera obscura* to draw outdoor images so that they would have correct visual perspective. By building this sketch box, you can use a magnifying glass to project a scene onto a piece of paper and trace its image.

What You Need
   magnifying glass (2-inch diameter or larger)
   tracing paper
   tape
   piece of glass, about 4 × 5 inches (available from
      a frame shop)
   2 cardboard boxes, as described in "What to
      do," below
   scissors or mat knife
   pencil
   black cloth
   silver duct tape

What to Do
   Find two cardboard boxes that can fit over one another rather tightly. See Figure 9. (Make sure that the length of the larger box is greater than the focal point of the magnifying glass you used.) Cut one end out of each box and slip the boxes over each other, with the cut ends

together. Now cut a rectangular hole (smaller than the plate of glass) from one box (box 1) and use silver duct tape to position the plate over this hole. Secure the black cloth on top of this box (the one with the glass plate) so that it acts as a light shield. Cut another hole in the end of the other box (box 2) just smaller than the diameter of the magnifying glass you used. Position and secure the lens over this hole.

Using the Camera Obscura

Place your completed camera obscura on a flat, sturdy surface. Use the black cloth as a hood, just as nineteenth-century photographers did when focusing their large studio portrait cameras. Move the boxes in and out until the large magnifying glass focuses a selected image on the glass plate. Is the image similar to the one observed earlier in your initial test? Tape a piece of tracing paper to the glass plate and trace the image. Remove the paper and transfer the tracing to another piece of paper by rubbing the tracing lines with the edge of a no. 3 pencil. Is this transferred image just like the original or is it a mirror image?

Experiment to see if you can form images without the use of a magnifying glass. (*Hint*: Replace the lens in the camera obscura by using a *very* small pinhole.)

ENGRAVINGS

Engraving is a method of carving a design or words on the surface of metal, wood, stone, or other hard material. The word "engrave" comes from a French word meaning "to carve in." The widest use of engraving is making plates from which to print; these prints are engravings. In making an engraving, the engraver uses a special tool called a *graver* (or burin) to cut into steel or copper plate by hand to plow out a V-shaped groove. This is an *engraving line*. After the engraving is complete, the whole surface of the

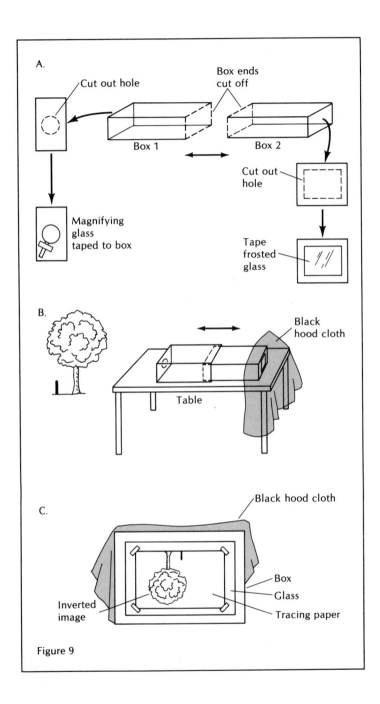

A.

Cut out hole

Box ends
cut off

Box 1

Box 2

Magnifying
glass
taped to box

Cut out
hole

Tape
frosted
glass

B.

Black
hood cloth

Table

C.

Black hood cloth

Box

Glass

Tracing paper

Inverted
image

Figure 9

printing block is covered with printer's ink and then wiped off. The ink remains only in the carved lines of letters or design. When paper is pressed against the surface of the plate, the ink in the groove is transferred to the paper.

Use the magnifying glass to examine various engravings such as paper currency or postage stamps:

• How many state names can you find engraved on the Lincoln Memorial pediment on the reverse side of a U.S. five-dollar bill?
• Are actual star shapes engraved in the U.S. flag that flies above the White House on the reverse of a U.S. twenty-dollar bill?
• How many individuals are riding in the automobile in front of the U.S. Treasury Building on the reverse of the U.S. ten-dollar bill?

Try making an engraving yourself. Use the point of a large nail to engrave a design or message on a flat piece of copper foil or copperplate (available at art supply stores). Use a large diameter (greater than 2 inches) magnifying glass while you work. Try duplicating some of the engraving designs you have studied. Spread a film of India ink over the completed engraving using a cotton applicator. Next, wipe off excess ink from the engraving plate using a

Making the camera obscura. (A) Box 2 fits snuggly so that it telescopes over Box 1. (B) Camera obscura is focused so that an enlarged and inverted image appears on the glass. (C) The inverted image can then be traced.

clean rag. Use a spray bottle to *slightly dampen* a piece of watercolor paper. Place the paper over the engraving. Carefully apply pressure over the surface of the paper so that the ink in the cut grooves will be picked up by the paper. (Use a wooden board to help apply uniform pressure to the engraving plate. Press down hard!) Remove the paper from the plate and allow to dry. Use the magnifying glass to examine the surface of your engraving print. You will notice that the inked line is above the surface of the paper. Look at a new sample of paper currency with your lens. Do you notice the same raised ink lines?

Use your magnifying glass to examine other types of engravings: look at inscribed jewelry pieces, silver utensils, glass; look for engraver's marks on these items as well.

IMAGES THAT COMMUNICATE

Computer bar codes are everywhere. They identify a particular product as well as provide pricing information that speeds check-out time. A harmless laser beam measures the space *between* the black bars, thus "reading" the numerically coded information.

Use a magnifying glass along with a black ink pen and a ruler to reduce white space between certain bar code lines. Also try adding additional lines to the bar code. With the permission of the store owner, see if your alterations can confuse the laser imaging system!

CAPTAIN EO!—THREE-DIMENSIONAL IMAGING

Many people have viewed this popular film in 3-D at Disney's Epcot Center. This effect is created by the use of the complimentary colors red (primary) and blue (secondary). Two picture images are made using these superimposed colors, and to the unaided eye present a blurred and confused appearance. When viewed through special glasses having a red (left eye) and blue (right eye) filter, each eye (depending upon the filter) sees only the picture intended for it, and the resultant images "fuse" to produce the three-dimensional image.

With practice, you can create your own 3-D images by photographing scenes in a special way and viewing these print images using a stereoscope. A stereoscope (available from sources listed in the Appendix) is an optical device made of two (2X) magnifying lenses mounted on wire supports. The lenses are separated by a distance (usually 2¹/₂ inches) corresponding to that between the eyes. The photo on page 40 illustrates an inexpensive stereoscope that can be purchased from supply companies listed in the Appendix. Use a stereoscope to view the stereoscopic image presented. Look through the stereoscope *directly* down at *one* selected common point. Your eyes should "fuse" these two distinct images into a single 3-D image.

Stereoscopic viewing was a popular activity in the mid-nineteenth and early twentieth centuries. Almost every affluent home had a stereoscope and a cabinet full of stereoscopic print images.

The key to making exciting 3-D images is careful photography followed by patience in arranging the print images during viewing.

What You Need
>   camera (black-and-white or color film; instant
>   films will also work)
>   ruler
>   sheet of white paper
>   pencil
>   stereoscope
>   utility blade
>   masking tape

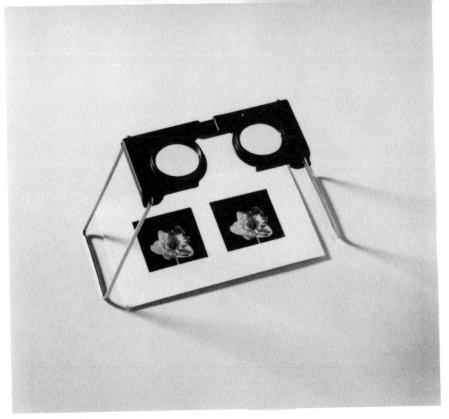

Stereoscopic imaging. Use a stereoscope (above) to view a stereoscopic image (below). Place the stereoscope over the center of these two images of the same daffodil flower. Look closely through both magnifying lenses of the stereoscope. Adjust them until the three-dimensional image appears.

What to Do

1. To make a 3-D image, you will take two separate photographs of the same subject. The first photograph is taken at a known starting point. The second photograph is taken *exactly* 3 inches (left or right from the center of the camera lens) from the original camera position. In both photographs the camera lens is aligned so that the same reference angle (90 degrees or perpendicular) to the base line is maintained. (If the reference angle changes for each photograph, the two resultant images cannot be fused by the brain.) Use Figure 10 as a guide in drawing the base and reference lines for camera positioning on a piece of white paper. Be sure to draw these lines carefully; use a ruler for accuracy.

2. Select a sturdy platform, such as a table, to support the camera. Position and tape the paper to the table surface. Make sure the paper is oriented correctly so that the two reference lines are at a 90-degree angle to the subject being photographed. For photograph 1, position the camera so that its back touches the base line and the lens centers on reference line 1 (left). Focus on the subject. Without moving the camera, make a number of exposures. Record the frame number along with "L" (left) for each exposure in your notebook.

3. Move the camera to the right along the base line. Position the center of the lens on reference line 2 (right). Repeat the exposures in the exact order for each set (left and right). Record the frame number along with "R" (right) as well as "L" (left). Have the roll of film developed.

4. Once the film is developed, use the magnifying glass to examine each frame image *in order.* Record

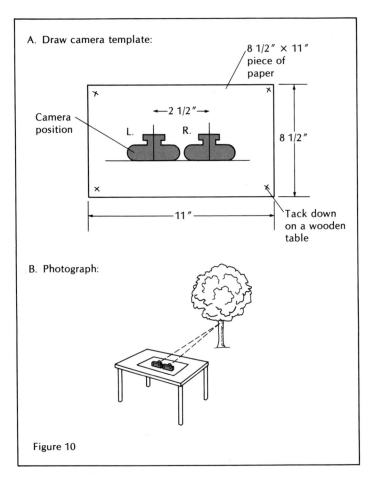

A. Draw camera template:

8 1/2" × 11"
piece of
paper

Camera
position

L.   R.

←2 1/2"→

8 1/2"

←——————11"——————→

Tack down
on a wooden
table

B. Photograph:

Figure 10

Making a stereoscopic image

both the frame number and the orientation (left/right) on the back of each photograph. Take a matched set of photographs and position as illustrated in Figure 10. Place the stereoscope over the picture set. The centers of each lens should be directly over a common point in both photographs. Use a ruler to check that the distance between this common point is approximately 3 inches. Adjust the stereoscope for best fit (correct interpupillary distance—about 65 mm) as you look *directly* down at *one* selected common point. Your eyes should "fuse" these two distinct images into a single 3-D image. If not, try moving each photograph slightly up or down until fusion occurs. A three-dimensional image is one that gives the impression of solidity and depth.

5. Any subject is a candidate for 3-D image viewing. After aligning the images using a stereoscope, make a permanent mount by taping the photographs to a card; trim ragged edges with a utility knife.

READ MORE ABOUT IT

Bolognese, D. *Printmaking.* New York: F. Watts, 1987.

Haines, G. *The Young Photographer's Handbook.* New York: Arco, 1984.

Hammond, John H. *The Camera Obscura: A Chronicle.* Bristol, England: Adam Hilger, 1981.

3

THE WORLD OF THE
LITTLE

A drop of water. Wet soil particles. Tiny roothairs of duckweed. What have they in common? They all are unique microworlds inhabited by tiny creatures awaiting discovery through the magnifying glass or bead lens.

This "world of the little" is all around us, and in us—anyplace that contains water. A spoonful of garden soil may contain more than 10 billion bacteria! These and other microlife forms dramatically affect our living world. (The term "microlife" applies to microscopic organisms such as protists, cyanobacteria, and fungi—life forms usually too small to be easily observed and studied with the naked eye.) Without microlife the earth's oxygen supply would be greatly reduced. Without microlife there would be few decomposers to recycle the dead living things whose decomposition renews the availability of nutrient materials for use by plants.

To view this inner-space world, you will need either a magnifying-glass microscope or a bead microscope. (See Chapter 1 for directions on building and using your own.) Use Figure 11 as your guide in identifying some of the microlife forms you will encounter.

LOOKING AT LIVING WATER

Why is pond or lake water sometimes green? Dip the eye of a canvas needle into collected water samples from a pond or lake and examine them with your microscope to find out. Also try sampling your home or school aquarium or a promising puddle. You should be able to observe protists with sizes of 300μ (0.012 inch) or larger, like *Paramecium* or *Volvox*. (The Greek "μ" is the symbol for micrometer. 1μ = $1/1,000,000$ of a meter or about $1/25,000$ of an inch.) Other, larger microinvertebrates—such as rotifers, ostracods, copepods and cladocerans (like *Daphnia*) —are always plentiful and easily observed through the lens. Perhaps you may encounter the behemoth *Bursaria,* big enough to see with the naked eye! Use Figure 11 as your visual guide to microlife.

GREEN BARK AND STONES

Protococcus is a green protist found growing as a green film on almost any surface outdoors—tree bark, rocks, and concrete. Do you find *Protococcus* growing mostly on surfaces facing a particular direction? Use a penknife to scrape off a sample, mix it with a little water, and examine with your microscope (Figure 11, no. 17).

Look for green rocks along the beds of streams. Again use your penknife to extract a sample to examine. Generally, these microlife forms are cyanobacteria (bacteria formerly called blue-green algae). They are long, blue-greenish strands, tufts, or tiny balls composed of many cells arranged end to end. Use Figure 11, no. 20–25 as a guide in identifying some of them.

MICROLIFE IN SOILS

This activity requires a bead-lens microscope.

Mix very small amounts of a collected soil type with

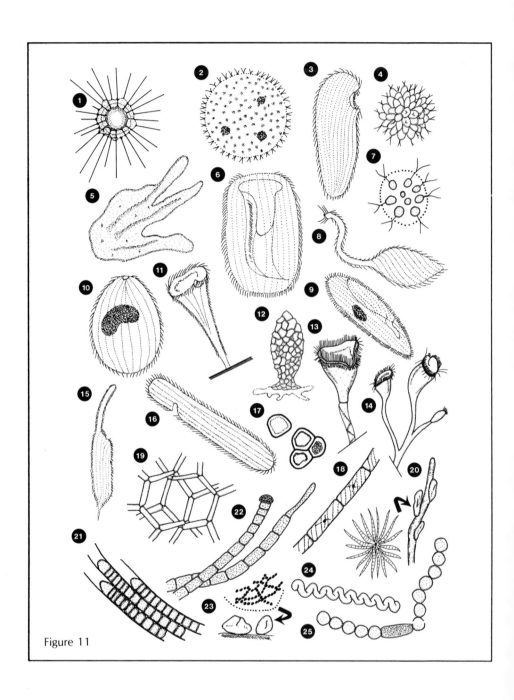

Figure 11

a number of drops of water in a paper cup. Try to capture individual grains with the eye of a canvas needle and examine them with your bead-lens microscope. Examine muds, wet beach sands, garden and forest soils. Use Figure 11, no. 3, 5, 8, 12, and 15 as a guide. Perhaps you will observe tiny transparent worms, soil nematodes.

Microlife forms. *Protists:*
(1) Actinosphaerium [300µm],
(2) Volvox [350–500µm],
(3) Loxodes [700µm],
(4) Synura [500µm],
(5) Chaos [1–5mm],
(6) Bursaria [0.5–1.0mm],
(7) Pleodorina [400µm],
(8) Lacrymaria [500–1200µm],
(9) Paramecium [180–300µm],
(10) Ichthyopthirius [100–1000µm],
(11) Stentor [1–2mm],
(12) Difflugia [230µm],
(13) Vorticella [150µm, cell],
(14) Carchesium [125µm, cell],
(15) Dileptus [250–500µm],
(16) Spirostomum [1–3mm],
(17) Protococcus [12µm, from bark scrapings], *(18) Spirogyra [macro], (19) Hydrodictyon [macro], (20) Cyanobacteria, (21) Gloetrichia* [macro], *Phormidium* [macro], *(22) Oscillatoria* [macro], *(23) Nostoc* [macro], *(24) Spirulina* [macro], *(25) Anabaena* [macro].

MICROLIFE ON OR AROUND AQUATIC PLANTS

Duckweed is a familiar aquatic plant found floating like a green carpet in the undisturbed shallows of most ponds and in roadside ditches. Place some duckweed (or a piece of another aquatic plant) in a specially constructed view box (see Chapter 1, Figure 7) and examine it with your magnifying-glass microscope. Use Figure 11 as a guide. Look closely at the surfaces of the aquatic plant. Can you observe and identify some of the same microlife forms *(Hydra, Carchesium,* and rotifers) that Leeuwenhoek did?

MICROLIFE IN INFUSIONS

Place a handful of finely chopped, dried grass clippings in a mayonnaise jar. Add distilled or bottled spring water (available at grocery stores) to the jar and allow to stand on a windowsill that receives northern light. Use the eye of the canvas needle to sample your infusion after about a week. Use Figure 11 as a guide to identifying microlife forms.

ICH

Ichthyopthirius is a parasitic protist that causes Ich disease on fish. The Ich disease can be observed as tiny

Macroinvertebrates, as they can be seen through the magnifying glass microscope. (A) Hydra. (B) Planaria. (C) Floscularia, a tube-building rotifer. (D) Soil nematodes, wriggling among garden soil particles.

white spots on the fins and body of an infected fish. Sample aquarium water where you observe this disease. Figure 11, no. 10 illustrates this large protist.

MICROFUNGI

Examine moldy objects with your magnifying glass: fruits, bread, and other decaying matter. Use the dichotomous (two-statement) key that follows to identify some of the fungi responsible for food spoilage.

KEY TO FUNGI THAT SPOIL FOOD

To use the key, begin at statement 1 and follow directions until you find the name of the fungus you are examining.

1a If there are fuzzy growth areas colored black, white, and/or gray, **go to 2.**

1b If there are fuzzy growth areas colored yellow or blue-green, **go to 4.**

2a If, under the magnifying glass, the hyphae (cottonlike mass) are not readily visible and there are numerous small spheres, the fungus is *Aspergillus flavus.*

2b If, under the magnifying glass, a white to gray cottonlike mass is seen, **go to 3.**

3a If the hyphae are beige-white and black spheres, the fungus is *Mucor.*

3b If the hyphae, gray to white and black spheres, are easily observed, the fungus is *Rhizopus.*

4a If the hyphae (cottonlike mass) are yellow with numerous dark spheres, the fungus is *Aspergillus niger.*

4b If the hyphae are blue-green to gray and few if any dark spheres are observed, the fungus is *Penicillium.*

Experiment with various foods to see whether any prevent spoilage or kill microfungi. Try vinegar, onion, garlic juice, salt, pepper, and lemon juice.

LANDFILLS: CONTAINMENT VESSELS, OR BIOLOGICAL RECYCLING CENTERS?

Today, available landfill space is at a premium. There is a growing demand for consumer products and product packaging that are biodegradable (capable of being broken down by microbes). With your magnifying glass try the following long-term activities to learn more about biodegradability.

STUDYING NATURE'S WAYS

In the fall, stake out a small (3 × 3 foot) plot in a secluded wooded area where leaves have fallen. Return to this marked plot at regular intervals in the following months (especially during early spring and summer) to find out what happens to the leaf litter. Can you find evidence of specific biological activity caused by microlife and animals? For example, does your magnifying glass reveal molds growing on surface and subsurface leaves? Do some leaves show evidence of being chewed by insects?

In addition to your notes and microscopic observations through your lens, take close-up photographs of the fallen leaves in the plot area at various time intervals, to help document your findings.

STUDYING MINIATURE "LANDFILLS"

Use transparent plastic cups to simulate "landfill" space. Select sample articles to evaluate: newsprint, cardboard (both thin and corrugated), leaves (both dry and fresh), grass clippings, "biodegradable" and regular plastic bags, and Styrofoam. Affix small pieces (1/2-inch square) of these refuse samples to 1-inch square pieces of clear acetate sheet (available at stationery stores) using glue or tape. Bury, at equal depths, a single sample in each landfill (plastic cup).

You can use different soils to simulate various environmental conditions. Test each article under the following conditions:

- *Dry soils.* Bake various soil types (topsoil, subsoil, sand, etc.) in an oven for about a half hour to remove water (under adult supervision).
- *Wet soils.* Use a spray bottle to simulate rain at regular intervals.
- *Landfill liner.* Seal your landfill with plastic wrap.

After an extended time period, (at least three or four months, preferably longer), remove your test articles and examine them with your magnifying glass. Record your observations (include drawings) under each set of environmental conditions:

- Do you observe evidence of biological degradation (microbial breakdown)? Is it similar to what you observed in your wooded test area?
- Is degradation faster in wet or dry soils? Why?
- Are all materials degraded; under all test conditions?

Can you draw any conclusion about biodegradability? Are newspapers and similar "degradable" articles being biologically degraded when placed in a landfill? Is a landfill simply a containment area in which little biological degradation is occurring? Would you expect biodegradability to occur in a landfill built in the Sahara Desert? Why or why not?

READ MORE ABOUT IT

Dobell, C. *Antony van Leeuwenhoek and His Little Animals.* New York: Dover, 1969.

Ford, B. *Single Lens.* New York: Harper & Row, 1985.

Margulis, L., and D. Sagan. *The Microcosmos Coloring Book.* New York: Harcourt Brace Jovanovich, 1988.

Rainis, K. *Nature Projects for Young Scientists.* New York: F. Watts, 1989.

Rolfe, R. T., and F. W. Rolfe. *The Romance of the Fungus World.* New York: Dover, 1974.

Sagan, D., and L. Margulis. *Garden of Microbial Delights: A Practical Guide to the Subvisible World.* New York: Harcourt Brace Jovanovich, 1988.

Stwertka, E. *The Microscope: How to Use It and Enjoy It.* Englewood Cliffs, N.J.: Julian Messner, 1988.

Ward's Dichotomous Key to the Protozoa. Rochester, N.Y.: Ward's Natural Science Establishment, 1985 (free publication).

4
THINGS ANIMAL

The magnifying glass allows you to see detail in the animal world your naked eye cannot observe. Through it, you will be able to better understand the intricate interweavings of nature's form and structure. Are insect joints and appendages the same as ours? How fast does a fish grow? Why do feathers float? Do earthworms walk?

The activities in this chapter, and other activities of your choosing, will help you to learn about things that are uniquely animal.

INSECTS AND INSECT STRUCTURES

About four-fifths of the million or so types of animals on Earth are insects. All insects have three pairs of legs, a body divided into three main parts—*head, thorax,* and *abdomen* —and a tough shell-like outer covering, or *exoskeleton.* Most insects also have wings, compound eyes, and a pair of antennae.

These fascinating animals do the same things we do,

54

but often in a very different way! For example, insects usually smell things through their antennae, and some taste with their feet. Many insects hear by means of specialized hairs *(setae)* on their bodies or on their legs. Some insects create noise by rubbing their legs together, a disturbance that sometimes can be heard for up to a mile! Insects breathe not with lungs but through external openings connected to internal tube systems. Some insects would easily be Olympic champions if they were our size. An ant can lift over fifty times its own body weight—no human can do that!

Insects are found almost anywhere. To capture most flying forms for study, use aerial nets, light traps (fluorescent light at night attracts certain insects; commercial light traps are available; see Appendix), or "sugaring." (Mix sugar with a little beer—under adult supervision—until you have a thick mixture. Apply the mixture with a brush to tree trunks at dusk. Return with a collecting jar and a flashlight to collect baited insects.) Use a dip net (a large fish-aquarium net) for aquatic insect forms.

Kill insects before study by using an insect-killing jar. You can make an insect-killing jar by using a small amount of nail polish remover (ethyl acetate) on a cotton ball and placing it inside a small glass jar that has a screw-top lid. **(*Caution:* Avoid inhaling any vapors!)**

Use a point holder (a sewing needle whose eye is embedded into a 6-inch length of 1/4-inch wooden dowel), a pin, or a pair of tweezers to hold the specimen for observation.

Use Figures 12 and 13 as a guide to external insect anatomy. Make drawings and take notes of your observations in your notebook. Use the field-guide books listed at the end of this chapter to identify your catches.

Insect skeletons. The skeleton of an insect is on the outside of its body and is made of a unique substance called *chitin,* which is lighter and stronger than bone.

Use your lens to examine the body areas of live insects. Can you identify the three main parts of an insect: head, thorax, and abdomen? Can you observe body hairs *(setae)*. Do these hairs occur singly or in groups? How is a body joint of an insect constructed?

Sometimes you can find an insect molt casing (insects periodically shed their exoskeleton) nearby (usually on tree trunks). Use your magnifying glass to study it. Are all body parts fused together, or are there flexible areas?

Eyes and antennae. Most insects have two large compound eyes made up of many different individual lenses called *facets.* All the lenses combine to form a complete picture of what an insect sees. A housefly has over 4,000 such facets that make up its eye; dragonflies have up to 28,000! Use the magnifying glass to study insect eyes; are compound eyes shaped the same? If possible, use your lens to look through the eye area in a molt casing. Can you see the way an insect does?

Almost all insects have two antennae between their eyes. They use their antennae chiefly to smell and to feel. Some insects also use antennae to taste and to hear. Compare antennae types on your specimens to those illustrated in Figure 12. How many types can you find? Do all have body hair? From your observations, can you suggest a specific use for a particular insect antenna?

Observe an ant colony with your magnifying glass. How do ants keep their antennae clean?

Mouthparts. There are two main types of insect mouthparts. One type is adapted for chewing, and the other for sucking. Each group of insects has its own variation of one of these types or both. Figure 12 illustrates some of them. See if you can identify the three main mouth-structure parts in an insect: *mandible, maxilla,* and *labrum.*

Only female mosquitoes bite warm-blooded animals to take a blood meal. Why don't male mosquitoes bite? *Hint:* Look for males around fruits and flowers.

Legs. Insects have three pairs of legs (six legs in all); each pair is connected to one of the three thorax segments. Use your magnifying glass to find out how many main segments all insect legs have. (A movable joint separates segments.) Observe insects with your lens. How do they walk? Are they always firmly balanced? On how many legs?

Many kinds of insects have legs adapted for special purposes. Figure 13 illustrates some of them. Use your lens to find out how flies walk upside down or up walls. How do honeybees collect pollen? How are aquatic insect legs modified for swimming?

Wings. Most adult insects have wings. Houseflies, mosquitoes, and all other true flies have two wings, which are attached to the middle segment of the thorax. Moths, butterflies, dragonflies, and other winged insects have four wings. One pair is attached to the middle segment of the thorax, and the other to the hind segment.

- Use your lens to observe as many different insect wings as possible. Are moth and butterfly wings alike? Are they similar to the wings of a housefly?
- How are insect wings supported—that is, what gives them the structural strength required for flight?
- Dragonflies and cockroaches are evolutionarily much older than houseflies or bees. Use your magnifying glass to observe rib (venation) patterns in wings in these insects. How do they differ from the wings of houseflies and bees?

SCALES TELL TALES

Scales are small, flattened plates forming a protective external body covering on certain animals: fishes, reptiles, the tails of a few mammals, the legs of certain birds, and parts of insects (butterfly and moth wing scales). Let's take a closer look!

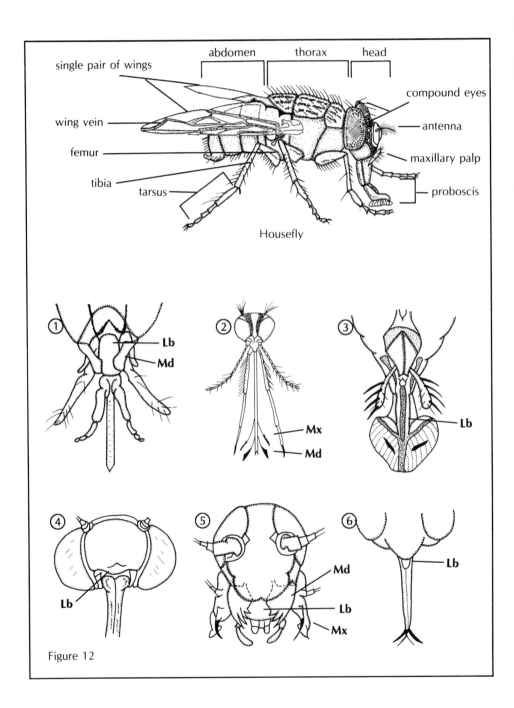

Figure 12

Insect scales. These small, flat structures are actually modified insect hairs! They make up the body and wing covering of moths and butterflies. Use fine-point tweezers or an artist's fine-bristle paintbrush to remove a scale for detailed study with your lens. Notice that body color of native species is the result of pigments embedded in the scale; exotic forms have iridescent colors due to surface sculpturing of the scale itself.

Reptile scales. Make a visit to a local zoo and ask the zookeeper for a piece of molted skin from a lizard or snake. Use your magnifying glass to examine it. Notice the small hinges between the larger scales. Why do you think they are necessary?

Fish scales. There are three principle types: *ganoid, cycloid,* and *ctenoid* (Figure 14). Only a small portion of the scale is exposed on the animal (like roof shingles); the rest is embedded within the lower skin. Ganoid scales are rhombic-shaped (an equilateral parallelogram) and occur in fishes such as pikes, gars, and sturgeons. They are attached to each other by joints. Because these scales are coated with a chemical called gamion, the fish have a polished, waxy appearance. Cycloid scales are usually circular, with concentric rings around a central point (focus), and are found in minnows, trout, and other soft-

Insect anatomy I. *Mouthpart types:*
(1) chewing-lapping (honeybee),
(2) piercing-sucking (mosquito),
(3) sponging (housefly),
(4) siphoning (butterfly),
(5) chewing (beetle),
(6) piercing-sucking (bug).
Lb=*labrum,* **Md**=*mandible,*
Mx=*maxilla.*

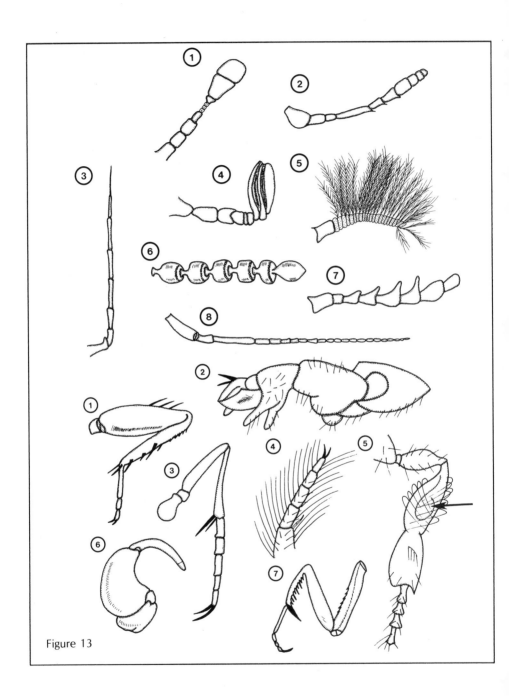

Figure 13

rayed fishes. Ctenoid scales are similar to cycloid scales except that the posterior (exposed end) has small spines or teeth. Spiny-rayed fish such as perch and sunfish have ctenoid scales.

Visit the supermarket or a fish wholesaler and ask the store manager for permission to obtain fish scales. Place them in marked envelopes (which you should supply), and bring them home. Soak the scales in clean water for about an hour, then scrub the surface with an old toothbrush. Use tweezers when handling scales. Place them on black construction paper and examine under direct lighting with your magnifying glass. Refer to Figure 14. Notice the many ringlike bony ridges *(circuli)* between the lighter areas called year marks *(annuli)*. These ridges are laid down during periods of growth. In certain scales (cycloid types) you may be able to see pigment areas (chromatophores).

Use white cement to attach scales to index cards. Label the cards with the name of the fish along with a small sketch of the scale as you observed it through your lens.

More insect anatomy. *Antennae Types:*
(1) with head (capitate), (2) clubbed
(clavate), (3) tapering (setaceous),
(4) leaflike (lamellate), (5) comblike
(pectinate), (6) beadlike (moniliform),
(7) sawlike (serrate),
(8) threadlike (filiform).
Leg Types: (1) jumping,
(2) digging, (3) walking,
(4) swimming, (5) pollen-carrying
[arrow = pollen ball],
(6) clasping, (7) grasping.

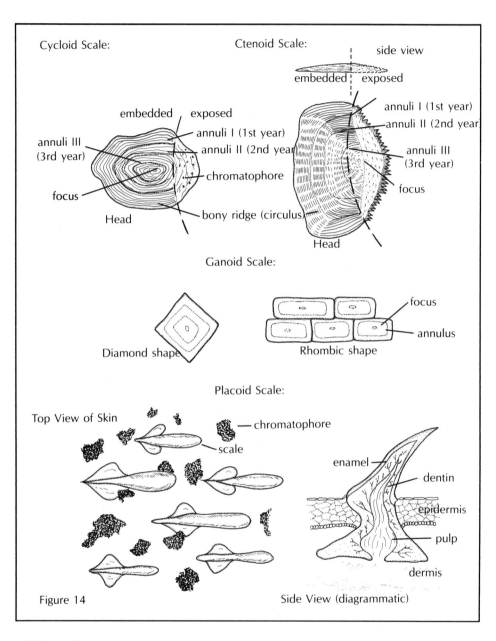

Cycloid Scale:

annuli III (3rd year)

focus

Head

embedded / exposed

annuli I (1st year)

annuli II (2nd year)

chromatophore

bony ridge (circulus)

Ctenoid Scale:

side view

embedded | exposed

annuli I (1st year)

annuli II (2nd year)

annuli III (3rd year)

focus

Head

Ganoid Scale:

Diamond shape

Rhombic shape

focus

annulus

Placoid Scale:

Top View of Skin

chromatophore

scale

enamel

dentin

epidermis

pulp

dermis

Figure 14

Side View (diagrammatic)

Fish-scale types

JAWS!

What do you and sharks have in common? Shark scales *(placoid scales)* are like your teeth—both have a cap of dentin that encloses a pulp cavity and is itself covered by hard enamel (see Figure 14). Fish markets usually carry shark steaks, which still has sharkskin attached. Run your finger over the skin. Does it feel like sandpaper? Use the magnifying glass to examine sharkskin closely. You will see the tiny placoid scales embedded in it. Use tweezers, while looking through your lens, to remove an individual shark scale for a closer look.

HOW OLD IS THAT FISH?

Often fisheries biologists determine catch limits and season length by finding the age of a given population of fish. Use Figure 14 as a guide when determining the age of fish scales. (The ctenoid scales of perch are best to begin with.)

First identify year marks (annuli). They will appear as lines around the embedded margin of the scale. When growth subsides in winter, a very small portion of the scale at the outer edge is reabsorbed by the fish. As growth begins again in the spring, it leaves a noticeable ring more obvious than the previous circuli. One annulus is formed each year.

Now examine the tiny bony ridges between annuli; they will reveal knowledge about growth periods and external factors (temperature, seasonal, and social) that affect it. If overcrowding, reduction of food, or unusually low temperature conditions exist, individual scale ridges (circuli) will be smaller and closer together. Look for "cut-offs," which indicate a check (cessation) in growth.

This method of age determination in fish is more accurate in the northern climates, where seasonal temperatures are more extreme. Biologists may also use fin spines to determine the age of fish.

By and large, few fishes live longer than twelve to twenty years. Usually fish that grow longer than a foot (30 cm) have a lifespan of at least four or five years. If possible, visit the market and ask for some sturgeon scales. How long do sturgeons live?

RADULA AND FEET

After insects, mollusks are the most diverse group of animals on Earth. A snail is an animal whose soft body is usually covered with a coiled shell. Snails creep along on a strong muscular organ called a *foot.* Look for aquatic snails that are moving along the side of an aquarium tank. Study the foot with your magnifying glass. How does the foot propel the snail?

Use your lens to study the mouth of the snail. You will see the *radula,* a toothed organ used by the snail to process the rich green protist lawns usually found on glass and plant surfaces in the aquarium.

EXAMINING ANIMAL HAIR

Human hair, and the hair of all mammals, grows from a *hair follicle,* a porelike organ within the skin. Hair is made up of two distinct parts: the *shaft,* or the portion that we see projecting from the skin, and the *root,* which is embedded in the follicle lying deep within the skin.

At any given time, your hair is in either the growing, resting, or dying stage. A hair of the human scalp usually grows about half an inch each month for two to four years. Then it falls out and a new hair replaces it. Use the magnifying glass to study your own hair. Hold a strand between your fingertips or use a pair of tweezers. Compare various strands to those illustrated in Figure 15. Can you observe actively growing and mature (dying) hairs?

- Can you identify hair from different individuals through color?

64

*This aquatic snail is exploring the side of an aquarium, using its radula (indicated by **arrow**) to scour algae deposits on the glass.*

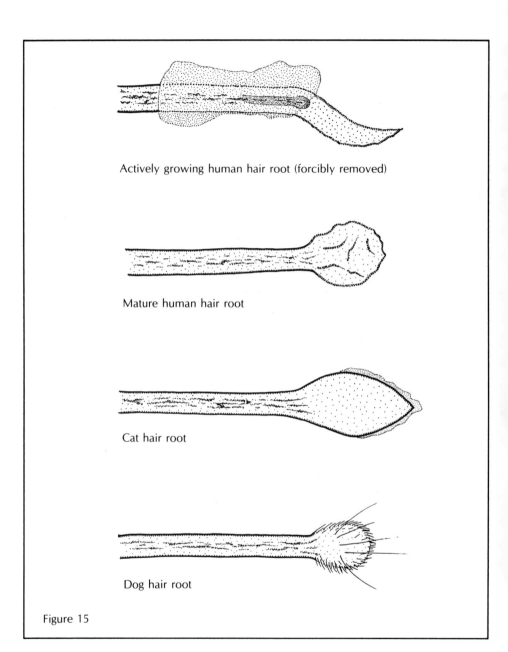

Actively growing human hair root (forcibly removed)

Mature human hair root

Cat hair root

Dog hair root

Figure 15

Animal hairs

- Use Figure 15 as a guide in identifying two other common animal hair types. Find other animal types and examine them. Make a collection of various hair types (hamster, guinea pig, gerbil, etc.) by placing these animal hairs in small labeled envelopes. In your notebook make drawings of the root ends of animal hairs. See if you or your friends can identify "unknowns" that are presented to them.

FEELING FOR HAIR SCALES

Try this experiment. Obtain a rather long strand of hair from your head and hold it by the root between your left forefinger and thumb. Exerting moderate to firm pressure, draw the hair strand, from the root to the tip, through the forefinger and thumb of your right hand. Now reverse the procedure, drawing the hair strand from tip to root. Which direction offers more resistance? The resistance you feel is due to the presence of thousands of cells that make up the outer layer of hair—the *cuticle*. These are laid down as overlapping scales with their edges pointing toward the tip. Generally you need a microscope (at least 50X) to observe cuticle scales.

- Use this touch method to examine other hair types, such as wool.
- If you live near a zoo that houses porcupines, ask the keeper for a hair sample. Do scale edges point frontward or backward? Why?

A NEW LOOK AT EARTHWORMS

Earthworms are nature's subterranean engineers. They are found in warm, moist soils throughout the world, feeding on dead plant matter. If you have a garden or lawn or live in the country, dig up an earthworm carefully. If you live in the city, you can get one from a bait or pet shop. Then try these activities:

- Gently run your finger over the bottom portion of the worm. Does it feel rough? Use your magnifying glass to observe sets of tiny hairlike structures (setae) on the lower portions of each body segment. How do earthworms use setae when moving about?
- Place an earthworm on a wet paper towel. Notice a prominent dark blood vessel on its top surface. Fold the paper towel over the worm, leaving a small window to observe this blood vessel near the head. Use the magnifying glass to observe blood flowing in this vessel through the window made in the paper towel. Record the beats per minute in your notebook. Now place a *thin* layer of crushed ice around the earthworm and record the beats per minute. What happens? Remove the ice. Slowly add drops of warm (but not hot) water. Again observe and record the beats per minute with your lens. Has a change occurred? Return your earthworm to the garden or lawn when you are finished observing.

FEATHERS

The feathers of birds help them fly and also keep them warm. Birds have two major types of feathers—contour and down. *Contour* feathers are the large feathers that cover the wings, body, and tail. Study Figure 16 to learn the parts of a contour feather. *Down* feathers are small, soft feathers found beneath the contour feathers of ducks, geese, and other waterfowl. Compare down feathers to contour feathers. You can obtain down feathers from certain winter clothing (usually a feather or two can be seen and pulled from a clothing seam; get permission before you do so). Contour feathers are easy to obtain; many clothing stores sell feathers.

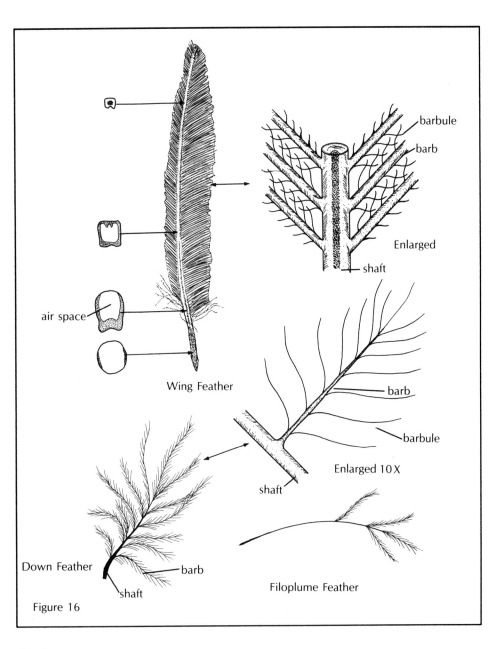

barbule

barb

Enlarged

shaft

air space

Wing Feather

barb

barbule

Enlarged 10 X

shaft

Down Feather

barb

shaft

Filoplume Feather

Figure 16

Feather anatomy

Choose a contour feather and hold the *quill* toward you. Examine it with the magnifying glass. Notice the small hole that allowed the entrance of blood vessels to nourish the feather during life. Obtain a large sewing needle and gently thrust it up into the shaft. What do you observe? Is this a benefit? Now take a single-edged razor blade and carefully shave the quill—like you are whittling a piece of wood. Examine these shavings with the magnifying glass. Compare them with fingernail cuttings from a nail clipper. Do they appear similar?

Try to spread one *vane* of the feather between your fingers (pull gently on the outer margin just below the top of the feather). Hold the feather up to a strong light and use your magnifying glass to examine the space between individual *barbs.* Note that these spaces are filled with minute structures called *barbules* which crisscross at opposite angles. Is it apparent from these observations that minute air spaces are present? How does the structure of the contour feather compare with a down feather? Do feathers float?

Separate individual barbs on the feather vane. Use indirect lighting to study individual barbules. Now run the feather vane back through your fingers (from the bottom toward the top of the feather). What happens? Instead of fingers, birds use their beaks, or bills, to realign displaced barbs. If a microscope is available, use it to examine an individual barbule. You will notice tiny hooklets. Hooklets are the secret to maintaining smooth, flexible vanes.

BIRDS AND OIL SPILLS

What happens if you coat feathers with a thin film of motor oil, simulating the conditions present in a major oil spill?

Use a detergent (like the special gel compound sold in stores to clean oily hands) to try to clean the feathers. Take notes on your procedure. Can you get barbs to hook themselves together? Are the minute air spaces, created by barbs and barbules, still present? Do these "cleaned" feathers float? Go to a library and research survival rates of birds whose oil-soaked feathers are cleaned by applying detergents or solvents. Do your experiments support these findings? Can you think of a way to rescue an oil-soaked bird?

A HISTORY OF TIME—FOSSILS

Fossils are traces of ancient life in rocks. Such traces may be tracks, trails, impressions, skeletons, shells, or fragments of remains. A good place to look for fossils is fossil beds—generally outcroppings of layered sedimentary rock, like shale or limestone.

At some time in the geological past, all the present land surfaces we now see were once covered by oceans. Animals and plants which lived in the sea accumulated on the bottom, to be subsequently buried and preserved. Sediments which settled on the sea bottom at that time have petrified (turned to stone). Over time, these deposits were uplifted by earth forces to form present-day mountaintops or grassy plains. For this reason, fossil remains of marine snails, clams, and even fish can be found in these current locations.

Often new fossil beds are revealed when roads are built or foundations for buildings are dug. Other places to find fossils include limestone quarries, and along river and stream beds. Be sure you have permission to collect in any area!

Use Figure 17 as a guide in identifying the more common fossil types. Use your magnifying lens to study individual rock or shale pieces. How can you tell if something you have found is truly a fossil? It must have a definite organic (life) structure. Do you live near a museum of natural history or a university? If so, the museum curator or university paleontologist (a scientist who stud-

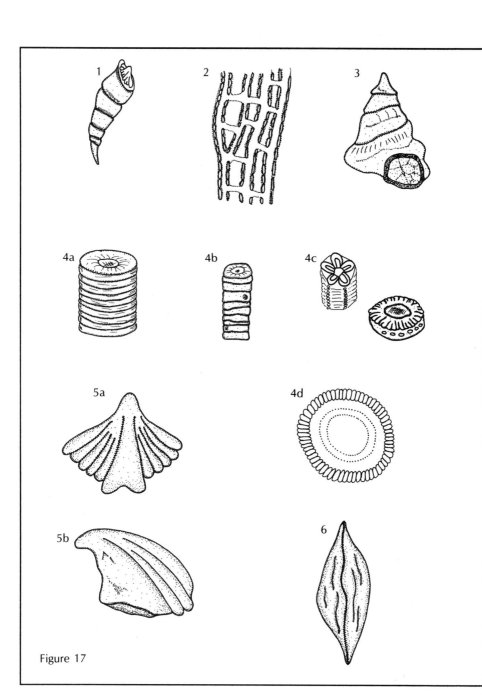

Figure 17

ies fossils) will be happy to show you their collections and help you identify the specimens in yours!

Search for more modern-day forms along the sea-shore. Often, by sifting through the sand you will be able to discover shell fragments, sea animals . . . maybe even a shark's tooth!

READ MORE ABOUT IT

Headstrom, R. *Adventures with Insects.* New York: Dover, 1982.

How to Make an Insect Collection. Catalog no. 32W2196. Rochester, N.Y.: Ward's Natural Science Establishment, 1978.

Pope, J. *Insects.* New York: F. Watts, 1984.

Rainis, K. *Nature Projects for Young Scientists.* New York: F. Watts, 1989.

Rhodes, F.; H. Zim; and P. Shaffer. *Fossils: A Guide to Prehistoric Life.* New York: Golden Press, 1962.

Thompson, I. *The Audubon Society Field Guide to North American Fossils.* New York: Knopf, 1982.

Fossil identification guide.
(1) Coral (common), (2) bryozoan or "moss animal" (common), (3) snail (very rare), (4 a-d) crinoid stems from extinct echinoderms (common), (5 a-b) brachiopod or lamp shell; top and side view (very rare), (6) foraminiferan (common fossil protozoan).

Winchester, A., and H. Jaques. *How to Know the Living Things*. Dubuque, Iowa: William C. Brown, 1981.

Zim, H., and H. Smith. *Reptiles and Amphibians*. New York: Golden Press, 1987.

———. *Insects*. New York: Golden Press, 1987.

5
THINGS VEGETABLE

In the last century, naturalists placed all living things into one of three groups: animal, vegetable, and mineral. This classification scheme has since been greatly revised and now includes five kingdoms of life—plant, animal, fungi, protist, and bacteria. Scientists group organisms in a particular kingdom because the organisms share certain basic characteristics and evolutionary relationships. These characteristics include physical form and structure, means of obtaining food, and means of reproduction.

Plants have a number of characteristics that set them apart from other living things. Plants make their own food (using the sun's energy, along with air and water) by a process called photosynthesis. Almost all plants stay in one place their entire lives. All plants develop from a tiny form of the plant called an embryo.

A plant is made up of several important parts. Flowering plants, the most common type of plants, have four main parts: (1) *roots,* (2) *stems,* (3) *leaves,* and (4) *flowers.* The magnifying glass will help you explore the world of plants.

LOOKING AT CELLS—"A GREAT MANY LITTLE BOXES"

Robert Hooke (1635–1703), an English scientist, was the first to use the microscope to probe and explain the physical properties of cork. Why was this natural material light, easily compressed, and water-repellent? In reporting to the Royal Society in April 1663, he coined the term *cell* to describe the structure of this unique plant tissue. Later, in 1665, he published illustrations of cork thin sections, describing them as "a great many little boxes," or cells, in his stunning work *Micrographia*.

See if you can repeat Hooke's observations using either a bead-lens or magnifier microscope. Use Figure 18 as a guide. Obtain a piece of cork from a cork stopper. Use a new single-edge razor blade to face it for fine cutting. Carefully shave *very thin* pieces off the original face cut. Make a number of sections. Select the *thinnest* one. Use tweezers to place a single cork section on a drop of water. Use the eye of a canvas needle to pick up the section held in the water drop. Observe it with your lens against a black background using a strong light as a point source.

Some of the largest plant cells known are those composing stamen hairs in spiderwort plants. Purchase a spiderwort plant that is in flower from a local florist or commercial grower. Use your magnifying glass to observe stamen hairs. (See Figure 19.) Pluck a stamen hair from a flower and observe a single cell through your lens. Each single bead is an individual plant cell. (You will be able to see the cell nucleus if you view a wet mount of stamen hair cells under a compound microscope.)

LEAF HOLES

Stomates are tiny openings in a leaf or stem through which gases pass. Most stomates are located on the un-

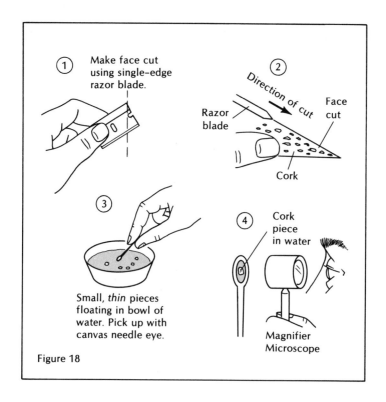

① Make face cut using single-edge razor blade.

② Direction of cut
Razor blade
Face cut
Cork

③ Small, *thin* pieces floating in bowl of water. Pick up with canvas needle eye.

④ Cork piece in water
Magnifier Microscope

Figure 18

Procedure for looking at Hooke's "great many little boxes": cells

dersides of leaves, and there may be thousands of them per square centimeter.

Make stomate impressions by painting a leaf surface with a thin layer of Duco cement. Allow to dry. Use fine-point tweezers to carefully separate the dried layer of cement from the leaf. Dip the cement cast into water and lay it on a clean microscope slide. Hold the slide up to the light (indoor incandescent or fluorescent lighting) to view

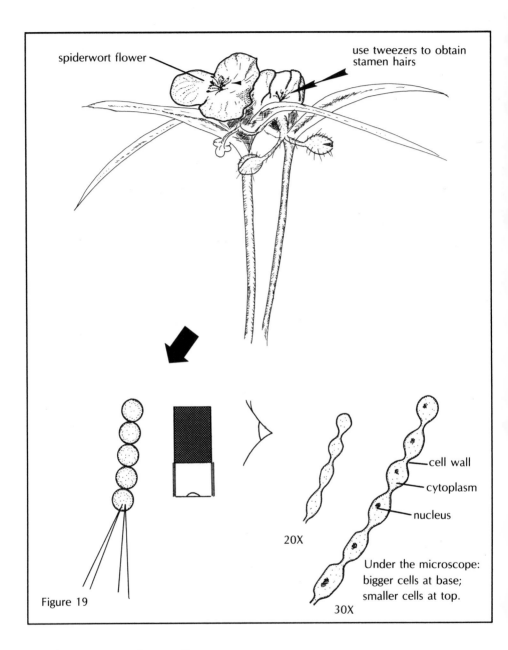

spiderwort flower

use tweezers to obtain
stamen hairs

cell wall

cytoplasm

nucleus

20X

30X

Under the microscope:
bigger cells at base;
smaller cells at top.

Figure 19

Studying stamen hair cells

the stomate casts with your lens. Although best viewed at 30X, if you look carefully (vary the lighting) you should be able to discern them. They will be *very small*! Use Figure 20 as a guide. (Stomates will be clearly visible under 40X power of a compound microscope. Make a wet mount to view them under this higher magnification.)

Make stomate casts and compare their number from leaves from the same tree or plant. Are they the same? Compare stomate numbers from casts made from different types of plants. Is there a difference?

WINTER TWIGS

How can you identify a deciduous tree in winter—without leaves? Use your magnifying glass to examine the twigs of any tree after its leaves have fallen. Look for light-colored areas, called *leaf scars,* formed when leaves fall off. Use your magnifying glass to examine a scar closely. You will observe a number of small, raised dots. These are *bundle scars*— the ends of vessels that carried water and nutrients to and away from the leaves. (Leaves are shed to protect the deciduous tree from excessive water loss during winter when moisture is usually retained deep underground in the roots.) The shape of the leaf scar as well as the arrangement and number of bundle scars are unique "fingerprints" that are used to identify a particular species.

Make detailed drawings, similar to the ones I have made in Figure 21, of winter twigs around your house and neighborhood. How many different shapes and arrangements can you find? Collect twig samples and label them using tags attached to the twig by string.

In summer, carefully break a leaf stalk off where it is attached to the twig. Try and match the vascular (vessel) bundles between these two structures. (Under the magnifying glass, these vascular bundles will look like tiny

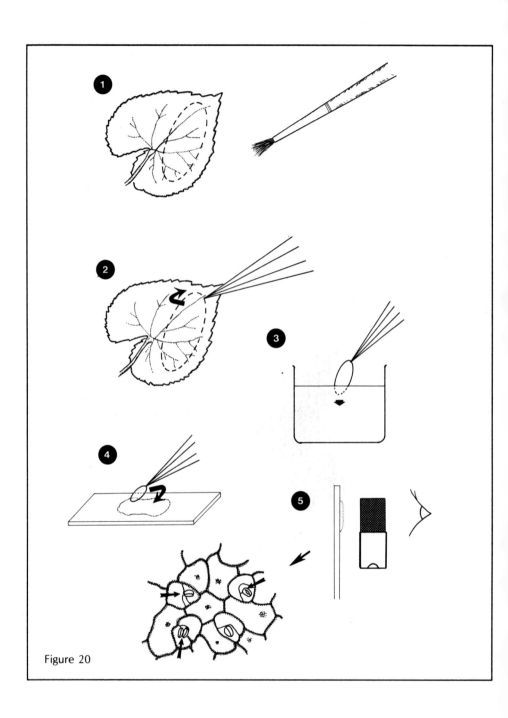

Figure 20

pipes. Celery stalks are great for the beginner!) To help you see vascular bundles more clearly, place leaf stems (or celery stalks) into water colored by blue or red food coloring. Place the leaf in direct sunlight. Use the magnifying glass to observe the dye migration up the stalk and into the leaf as well as at its attachment point between stalk and branch.

GRASS IS NOT ALWAYS GREENER ON THE OTHER SIDE

Today, many homeowners contract with lawn-care services that use a liquid fertilizer to provide for summer-long green lawns. In addition to making seed, grasses usually multiply by the formation of *rhizomes* (a modified underground stem). To observe rhizomes, use a spade or trowel to dig up a small square piece of lawn (first ask your parent's permission). Gently wash the root system with water. Use your magnifying glass to observe the interconnecting root systems (rhizomes) of the various grass plants.

Locate a yard whose homeowner is regularly using a

Observing leaf stomates:
(1) Brush with Duco cement.
(2) Pick up dried cement cast from leaf with tweezers.
(3) Dip leaf cast into water. (4) Place leaf cast on microscope slide.
*(5) Observe stomate impressions (**arrows**) against a lighted background with either a magnifying glass or a microscope.*

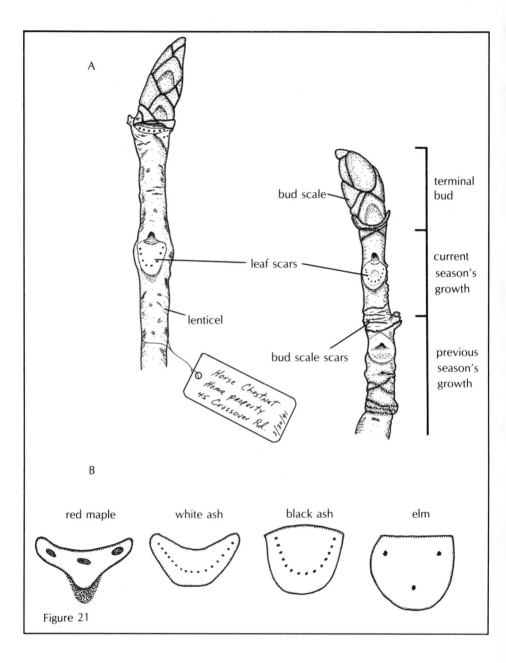

A

bud scale

terminal bud

leaf scars

current season's growth

lenticel

Horse Chestnut
Home property
45 Crossover Rd. 3/30/91

bud scale scars

previous season's growth

B

red maple

white ash

black ash

elm

Figure 21

lawn-care service that applies a liquid fertilizer. (Or ask your parent's permission to use a liquid fertilizer to regularly treat a small area of your lawn.) Ask permission to dig up a small (3 inch square) piece of turf. Compare rhizome formation in turf with and without application of liquid fertilizers. What effect does the regular application of such fertilizers have on rhizome development? (You can also try varying the amount of liquid fertilizer, by diluting it, as well as the frequency of application. Use a separate test plot for each type of fertilizer application.) How close to the surface are rhizomes in turf with regular liquid fertilizer applications as compared to turf in which no such application was made? In most locales, in late summer a lack of rain makes lawns turn brown. Do fertilizer-treated lawns (that are not watered) turn brown quicker than nontreated lawns? What conclusions can you draw regarding the benefit(s) of such lawn-care practices?

BUDS: A STUDY OF PROGRAMMED GROWTH

You can identify a tree in winter by its buds. Buds found on branches of trees and shrubs in winter were formed the preceding summer. Each year when a leaf bud grows in the spring it usually forms a new branch on the end of an old one. This new part of a tree's branch bears leaves. However, it will not bear leaves the following year, when new buds grow.

*Winter twig bundle scars.
(A) Winter twig anatomy in
horse chestnut. (B) Leaf
scars of some common trees.*

Buds that grow at the ends of stems are called *terminal buds*. Buds which grow on the sides of twigs are called *lateral buds*. These lateral buds are also formed at the base of each leaf stalk. They are also called *axilliary buds* (Figure 21). Other lateral buds which grow anywhere on the plant except at the leaf stalk are called *adventitious buds*. They usually grow as a result of injury, and may appear on roots, stems, and leaves. Can you find examples of adventitious buds? *(Hint:* Partially submerge a cut potato, suspended by toothpicks, in a glass of water.)

Use your magnifying glass to examine the outer surfaces of buds. You will observe that buds are made up of outer modified leaves called *scales,* which cover the delicate young leaves, stems, and in some cases flowers inside. Leaf buds of woody plants have tough scales to protect these inner structures against cold winters. As you examine various bud scales, you will notice a great deal of variation. Some scales are covered with fine hair (willow, sugar maple); others have a waxy covering (cottonwood) that serve to protect internal structures. How many variations of leaf scales can you find?

INSIDE INSECT HOUSING PROJECTS—GALLS

A gall is an unnatural growth on a plant that is caused by an insect larva (sometimes by worms, fungi, or bacteria). Galls may be observed on roots, stems, leaves, or even a flower bud.

Insect galls have many shapes. Some are smooth and round, like tiny balls; others are rough and hairy; some are shaped like pine cones. A gall forms after an insect has deposited an egg inside plant tissue. Following the emergence of the insect larva from the egg, the plant tissue is stimulated (in some mysterious way) to grow more rapidly and enclose the insect.

You can hunt for galls almost anywhere, in any sea-

Observing insect galls. This oak gall served as the home for seven developing oak wasp larvae of which one is visible (inset, center).

son. Look for insect inhabitants in early spring before they emerge. The best place to observe them is in fields with goldenrod or similar high-growing weeds. Also check marsh areas. Cattails are another excellent host plant. Around the home, look to rose bushes and willow and oak trees for galls. Pinecone-shaped galls on blue spruce trees are common as well.

When you find a gall, use your magnifying glass to observe its outer surface. Can you find evidence that the insect has vacated its protective shelter, or has a bird or squirrel already been there before you? Use a sharp penknife or single-edge razor blade to section the gall lengthwise. Use your magnifying glass to examine the insect larva housed within.

SEEDS: NATURE'S LIFE CAPSULES

In plants, seeds are produced so that future generations may continue to prosper. Some plants (mosses and ferns) do not produce seed. They reproduce by means of spores. Look for moss and moss capsules in wooded and marshy areas. Use your lens to examine moss spore casings. Carefully dissect a capsule by sectioning it with a single-edge razor blade.

A *seed* has three important parts: a protective outer skin, or *seed coat;* an *embryo,* which will become the new plant; and a food supply, or *endosperm.* Use a single-edge razor blade to section corn and lima bean seeds (lengthwise).

Gymnosperms are plants that produce "naked seeds." *Angiosperms* produce "enveloped seeds" (enclosed within an ovary which ripens as a pod or fleshy fruit). Naked seeds are those from pine and other cone-bearing trees. The seeds of all plants—trees, shrubs, and herbs—that bear flowers are angiosperms.

Use your magnifying glass to examine each of these

seed types. Which seed types have the largest embryos? The smallest? Refer to Figure 22 as a guide in identifying internal structures.

FLOWERS: NATURE'S BLOOM

The word "flower" may mean either the blossom or the whole plant. Botanists (scientists who study plants) use the word to mean only the blossom of a plant. They call the whole plant—blossom, stem, leaves, and roots—a *flowering plant.* Any plant that produces some sort of flower, even a tiny colorless one, is a flowering plant. Grasses, roses, lilies, apple trees, and oaks are all flowering plants.

Go to a florist and obtain either a daffodil, tulip, or gladiolus (depending on the season). Each will serve as a "typical" blossom for you to dissect and study with your magnifying glass. Use Figure 23 as a guide to flower anatomy. (Remember that sepals and petals are actually modified leaves.)

- Is there really anything like a "typical" flower?
- Are the tassels of corn really flowers? How can you tell?
- Does lawn grass make flowers? (*Hint:* Let some of it grow!)
- Observe insects on flowers. How do certain types of insects pollinate certain flowers?

JACK AND HIS BEANSTALK

Use your magnifying glass to observe how ivies (ornamental climbing vines) attach themselves to masonry or use other plants to get ahead. Do not investigate poison ivy, however (remember the old saying "leaflets three, let it be").

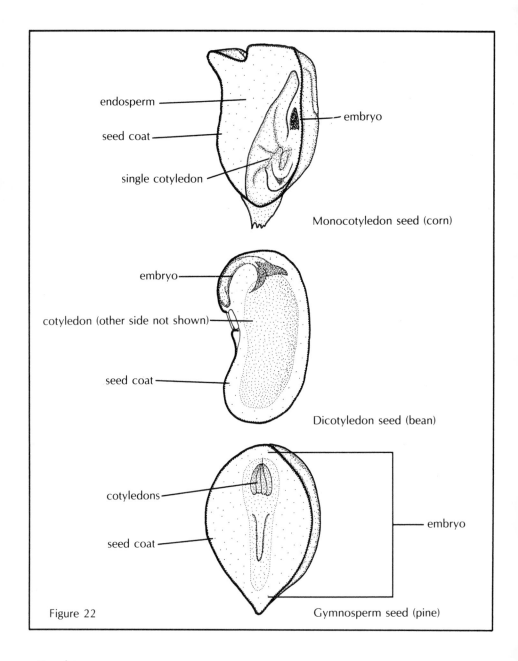

endosperm

seed coat

single cotyledon

embryo

Monocotyledon seed (corn)

embryo

cotyledon (other side not shown)

seed coat

Dicotyledon seed (bean)

cotyledons

seed coat

embryo

Figure 22

Gymnosperm seed (pine)

Seed types

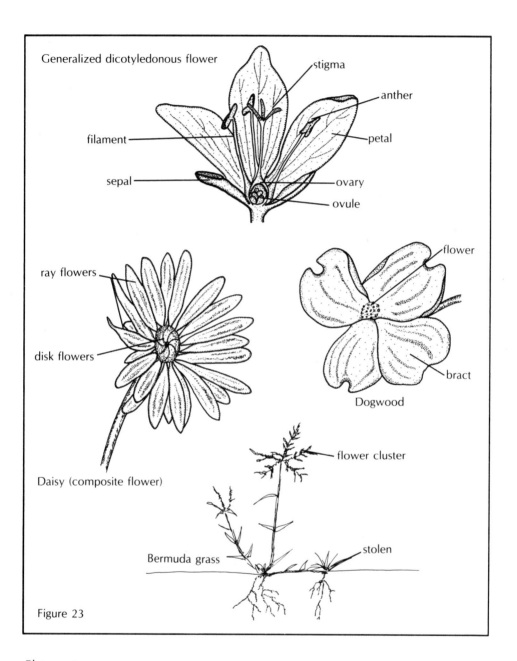

Generalized dicotyledonous flower

stigma

anther

filament

petal

sepal

ovary

ovule

ray flowers

flower

disk flowers

bract

Dogwood

Daisy (composite flower)

flower cluster

Bermuda grass

stolen

Figure 23

Flower types

TOBACCO, POLLEN, AND STARCH GRAINS

Use your magnifying glass to observe the special macroscopic detail of these familiar botanical objects:

Tobacco. Many large shopping malls still have tobacco shops. Go there and ask for small samples of the tobaccos offered for sale. Examine cigarette, pipe, and cigar tobaccos with your lens by placing them on a white card under direct lighting. Can you distinguish various types, or "blends," within a particular brand name? Is cigarette tobacco of a smaller size than pipe or cigar tobacco? While you are at it, examine cigarette filters as well.

The great fictional detective Sherlock Holmes wrote a paper on the subject of tobacco ash. If possible, use your lens to examine cigar, pipe, and tobacco ash on black construction paper. Can you really identify a particular brand, or type?

Pollen is produced in the *anthers* of flowers. Use tweezers to remove an anther and gently rub it on the surface of black construction paper. Examine this "smear" with your lens. Are pollen grains all the same size? Color?

Starch grains are easily obtained from potatoes. Place a small piece of potato in a cloth and press out a small amount of juice. Place a drop of potato juice on a glass slide. Allow to dry. Place the slide on top of black construction paper and observe under direct lighting with your hand lens.

READ MORE ABOUT IT

Beller, J. *Experimenting with Plants.* New York: Arco, 1985.

Hooke, R. *Micrographia* (1665). Facsimile edition. Lincolnwood, Ill.: Science Heritage, 1987.

MacFarlane, R. B. *Collecting and Preserving Plants for Science and Pleasure.* New York: Arco, 1985.

Petrides, L. A. *A Field Guide to Trees and Shrubs.* Boston: Houghton Mifflin, 1975.

Rainis, K. *Nature Projects for Young Scientists.* New York: F. Watts, 1989.

Winchester, A., and H. Jaques. *How to Know the Living Things.* Dubuque, Iowa: William C. Brown, 1981.

6

THINGS MINERAL

Unlike plants and animals, which are made of various parts, minerals are *homogeneous*—that is, every part is like every other. Minerals are not part of the living process, but of the physical, tied directly to the ability of the Earth to renew itself.

The magnifying glass will help you understand that minerals are highly organized crystalline solids (or crystals) that are unique in nature!

LOOKING AT "CLEAR ICE"—CRYSTALS

The word "crystal" comes from the Greek root meaning "clear ice." Long ago it was believed that glaciers were made of quartz. Today, you and I know they are simply mountains of ice and snow. Both ice (or snow) and quartz are solids; both are crystals. Crystals are highly organized solids having smooth, flat surfaces that meet in sharp edges or corners. Most nonliving things are made of crystals. Crystals are made from the outside in a process termed *crystallization*—adding layers of their own substance to what already exists.

Use your magnifying glass to examine table salt (sodium chloride) crystals under direct lighting by sprinkling some onto a piece of black construction paper. What geometric shape do you see? In your notebook, draw a three-dimensional illustration of a sodium chloride crystal. Note that your observations agree with the definition of a crystal—it is a solid having smooth, flat surfaces that meet in sharp edges or corners. Is it a "highly organized solid"? It is, but this organization is really at the atomic level—a salt crystal is made up of closely bound groups of sodium and chloride ions that together form what you see under your lens. This orderly atomic arrangement can be demonstrated for the magnifying glass using soap bubbles.

Place a clear glass dish containing soapy water on top of black construction paper. Insert a soda straw into the liquid and blow very lightly to create a single area of bubbles (a "bubble raft"). Look at the bubble raft with your magnifying glass. Notice that each bubble touches an equal number of other bubbles. How many?

Just as with soap bubbles, it is this orderly arrangement of atoms that forms a building block, a unit that repeats to make a highly organized solid called a crystal.

Based upon this knowledge, would you say that glass is a crystal? Use your magnifying glass and the reference books listed at the end of this chapter to find out.

GROWING MACROCRYSTALS

To grow a large crystal you need a beginning or "seed" crystal. Dissolve table salt in a glass of cold water. In your notebook, keep track of the amount (in teaspoons) that you use before the salt will not further dissolve, even after continuous stirring. (The resultant solution is called a *satu-*

rated solution—at a certain temperature no additional salt will dissolve in the water.) Now dissolve more salt in another glass that contains very hot water. (You will find that at warmer temperatures more salt can be dissolved in the same volume of water—this type of solution is called *supersaturated.*) Pour the clear salt solutions from both glasses into two other glasses. Cover both glasses with a handkerchief and secure the covers with a rubber band. Poke a small hole through the cloth in the center of each glass. Lay a pencil across the top of each glass. From each pencil let a piece of black sewing thread (knotted at the end) hang down through the handkerchief and into the water. Leave both glasses undisturbed in an area of your house where the temperature will remain constant over long periods (like the basement).

After a month or more, small cube-shaped crystals (called "halite") will have appeared on the strings. Examine your crystals with a magnifying glass. Do they all have the same shape as the table salt crystals you first looked at? Do some appear to have grown together? Mineralogists call crystals which have grown together and thus do not have flat faces "irregular" or *anhedral.* Crystals that are "regular," having natural faces, are *euhedral.* Crystals having only partly developed crystal faces are *subhedral.* What types of crystals do you observe?

Which solution (the cold *saturated* or the hot *supersaturated*) was best for growing the salt crystals? Which solution produced bigger crystals?

- Try growing crystals of other common substances: sugar, alum (aluminum ammonium sulfate, available in grocery stores).
- Obtain the following materials and use them to grow a macrocrystal garden. (Make sure you have your parent's permission before beginning. Wear safety goggles when adding these chemicals.) Obtain a

deep glass dish and in it place one or two pieces of brick (about 1 inch in diameter). Add the following around, but not on top of, the brick pieces so they are not covered:

6 tablespoons of water
6 tablespoons of table salt
6 tablespoons of laundry bluing
2 tablespoons of ammonia
2 tablespoons of dye coloring (any color)

• Place the bowl in a quiet area, and observe with the magnifying glass after two or three days.

GROWING MICROCRYSTALS

Follow the same procedure as you did for macrocrystals, but with the following changes. Once saturated and supersaturated solutions are made, use an eyedropper to place one or two drops of each on a separate clean microscope slide (or a piece of glass). Allow the water to evaporate so that the process of recrystallization can occur. Once all the water has evaporated, hold the slide, at a slight angle, over a piece of black construction paper underneath a table lamp. By holding the glass slide at an angle you will be able to more clearly see the intricate details of these crystals without light reflection. Observe the crystals with a magnifying glass (10X or higher magnification).

• Microcrystals of many interesting substances can be examined in this manner: aspirin, Epsom salts, vitamins, baking soda, lemon juice, even human tears! Sketch these shapes in your notebook. Keep your collection for future reference.
• You can observe ice crystals at any time of year! Breathe on a small piece of glass to form a film. Wipe off one side of the glass with a paper towel. Place the glass in a freezer (frost-free). Wait about

a day. Examine over a piece of black construction paper—be quick, before the crystals melt!

If you live in a northern climate, you can usually observe ice crystals on the inside of windows during winter months without going outside. If you do journey outdoors, take a piece of black construction paper or a piece of black velvet (prechilled in a freezer). "Catch" snowflakes as they fall and examine directly with your lens. Use your magnifying glass to find out how many sides any snow crystal has.

FIND-A-MINERAL

A *mineral* is a single substance in a specific crystalline form. A *rock* is made of one mineral or a combination of minerals. Minerals generally occur as individual small grains, while rocks represent a physical mixture of mineral grains. Granite is a rock that contains three basic minerals: quartz, feldspar, and mica. Marble is a rock made of cemented grains of a single mineral, calcite.

Use a large-diameter magnifying glass to examine granite, a material commonly used for roadside curbing and gravestones. What you see under magnification are aggregates of three (or more) minerals tightly compressed together—individual or whole crystals are not usually evident. This is because this is an *igneous rock,* formed by the crystallization of molten minerals as a result of volcanic activity.

So how do you identify the minerals present? One objective test is hardness, a basic property of minerals. *Hardness* is the resistance of a mineral to being scratched. By scratching a mineral with simple tools, its hardness can be compared to that of the tool. The Mohs' scale (Table 1) permits comparison of the hardness of minerals. Look at the scale. Are diamonds the only substance that cuts glass? What about jewelry made from quartz?

Table 1. Mohs' Scale		
Mohs' Number	Mineral	Comparison Object
1	Talc	Softer than fingernail
2	Gypsum	Softer than fingernail
3	Calcite	Harder than fingernail
4	Fluorite	Harder than fingernail
5	Apatite	As hard as an iron nail
6	Feldspar	Harder than window glass
7	Quartz	As hard as a steel file
8	Topaz	Harder than a steel file
9	Corundrum	Harder than a steel file
10	Diamond	Hardest known substance

Granite is a light-colored rock composed mostly of quartz and feldspar. Quartz crystals will have the appearance of broken glass. The quartz crystal does not *cleave* (break along a smooth, even surface); instead it *fractures* (breaks unevenly). A steel file will scratch quartz. If possible, use a small steel file to try to scratch crystals you suspect to be quartz. Use the magnifying lens to observe markings. (See "Severing Atomic Bonds," at the end of this chapter, to learn more about cleavage and fracture.)

Feldspar varies from light pink to grayish or white in most granites. It is softer than quartz, so your file scratches will be easier to make and observe.

Most granites are the "salt and pepper" variety. Usually the small dark-colored specks are mica, but they might be amphiboles, minerals that are rich in iron or magnesium. Use the hardness chart and your lens to find out. Mica is a relatively soft mineral (softer than amphibole), so a fingernail will easily scratch its surface. It also cleaves easily into fine, small plates. Once you have determined that the mineral is mica, its color will tell you whether it is muscovite (light cream) or biotite (dark).

Another common rock is marble, composed of a single mineral, calcite (calcium carbonate). Marble is a rock formed (recrystallized) as a result of heat and pressure applied long ago to preexisting rocks. We call these altered rocks *metamorphic*. Besides calcite, some marbles contain iron and other minerals, which can give it beautiful color patterns. In older sections of some cemeteries, white marble was commonly used as the stone for a grave marker. The white calcite crystals that compose these marbles have a much different appearance than either quartz or feldspar—they are often uniformly coarse-grained. Calcite is harder than your fingernail, but will not be able to scratch window glass. (Hardness points —small pieces of cut minerals mounted to the ends of brass rods—corresponding to the Mohs' scale, can be used to help identify the minerals you find. Mineral collections, as well as hardness points, can be purchased from suppliers listed in the Appendix.)

In Italy (and many other places) statues made of white marble are fast "disappearing." To find out why, go on to the next activity!

EFFERVESCENT MINERALS

Calcite is made of calcium carbonate. When this mineral reacts with an acid, it forms carbon dioxide. Obtain a small flake of white marble. (Ask the owner of a gravestone-monument or landscaping company for a small, waste piece of white marble, or calcite.) **Never alter or remove any material from headstones or other building materials.** Wear safety goggles. Place the small flake in a paper cup and add a small amount of white vinegar (5–7 percent acetic acid). Use your magnifying glass to observe tiny bubbles of carbon dioxide gas that evolve. Does this experiment help explain why certain art treasures are in jeopardy from industrial air pollutants and acid rain?

INVESTIGATING METAL FATIGUE

Engineers use the term "metal fatigue" to describe certain changes that occur in substances like metals. Repeated stress on a metal, such as continued weight, tension, bending, or pressure, may cause a progressive weakening. Such stress often alters the molecular structure of the metal (which has a crystalline structure) so that it bends or cracks.

A paper clip is an ideal object to use to investigate metal fatigue. Paper clips are made from carbon steel which is drawn through a machine into spools of wire. This wire is then cut and bent to its familiar form. Take a paper clip and open it up. Examine the middle of the wire with your magnifying glass. Write down your observations in your notebook. Bend the clip at the middle once in both directions. Again, look at the area you have bent and record your observations. Keep bending the wire at the same point in both directions, recording your observations after each full bend, until it breaks. (With some paper clips you will notice a flaking. This is a coating of another metal.)

Metal fatigue usually begins at the surface of a metal, where small defects serve as a concentration point for stress.

Here are some other activities you can try to investigate metal fatigue:

- Try making a mark in the paper clip with a file before you start to bend it. Does this mark promote or relieve stress that leads to breakage?
- Now try twisting another paper clip instead of bending. Record your observations. What is different about this form of stress? Try alternate bending and twisting.
- Compare metal fatigue characteristics in different metals: aluminum in gutters or rolled aluminum sheets, copper, lead. Are there differences?

- Secure a heavy strip of aluminum (or other metal), purchased at a hardware store, on the surface of a workbench so that one end extends away from the bench. Gradually add weights to the unsecured end. Find out how much weight causes the metal strip to bend. Afterward, examine the strip with your magnifying glass for signs of stress.
- Are tubular forms of metals more or less resistant to fatigue than sheet metals? Try copper pipe (compare the rigid to the soft, "flexible" type).

INVESTIGATING ALLOYS: SOLDERING

An *alloy* is made by heating two (or more) metals together.

Solder is a metal alloy used to join two metals together. To be effective, the solder must melt more easily than the metals to which it is applied. Solder acts only as an adhesive (like glue) and does not form a continuous, or welded, bond with the metals being joined.

Use your magnifying glass to observe a solder joint. If this is not possible, ask an adult to solder two pieces of copper pipe together. Work with a sample solder joint (**not a real one** currently in use in your house). Use a hacksaw to cut through it and observe with your lens how the alloy binds to both metal pieces.

There are two kinds of solder, hard and soft. Soft solder usually contains alloys of tin and lead. (Today, less lead is used in this type of solder in plumbing joints to minimize contamination of drinking water.) The most common hard solder is silver solder, which consists of silver, copper, and zinc. Go to a hardware store and ask for some samples of both solder types. Perform metal fatigue tests on each solder sample. Besides melting at a much higher temperature than soft solder, what other advantage does hard solder have?

A METAL'S SCOURGE—RUST!

Iron oxide (rust) is the brownish-red substance that forms on the surface of iron or steel when exposed to damp air. Rust forms by the union of oxygen (in air) with iron in metal. This process is called *oxidation*. Moisture is an important agent in producing this change.

Use your magnifying glass to examine rusty metal objects. Is rust a crystal? Use a wire brush to remove the rust. Is the metal itself corroded? Would rust-encrusted metals become weaker following prolonged exposure to the elements? Test commercial antirust compounds sold in hardware stores. Do they all work? Read the product labels to determine what compounds contained in these products are the "active ingredients." Do the commercial products that seem to work better have more of a particular compound, or a unique compound?

Aluminum is prized for its low corrosion rate. This is because aluminum forms a tightly adhering surface film of aluminum oxide when exposed to the air. Under most atmospheric conditions this layer is enough to protect it from corrosion.

Take several pieces of aluminum flashing (available at hardware stores) and allow them to "weather" for a month or two. Examine the surface with your lens for oxide formation by using a nail point to scrape a tiny groove in the weathered aluminum. Now immerse some of the weathered aluminum strips in a bucket of warm salt water. Leave them for a day or so. Use your magnifying glass to observe what happens. Would you side a house with aluminum sheeting if it were located near the seashore? Try other metals, such as copper. Copper forms a green patina (cupric oxide) that helps protect it against corrosion. See if you can develop a treatment (such as painting) that will allow aluminum to resist corrosion in a moist and salty environment.

LOOKING AT COINS

Metal currency is made by stamping an impression into metal "blanks" using a die in a stamping machine. A mint is the place where metal currency is made. Usually coins are minted by authority of a government.

Use your magnifying glass to examine metal coins. Do you notice variation among similar denominations? Is the depth of the stamping always the same? Is the quarter a single metal, an alloy, or a composite (layered metals)? Is a penny made of copper? (Use tin snips to **carefully** cut coins in half for further examination.) Use your magnifying glass to observe whether a penny corrodes in the same way that a piece of copper does (i.e., give it the "saltwater test").

There are also private mints in this country. They usually design special commemorative medallions stamped in precious metals. Some private mints even coin money under contract to foreign governments. If possible, use your lens to examine examples. Some have intricate and beautiful designs!

INVESTIGATING A WONDER OF NATURE—THE DOUBLE IMAGE

Iceland spar (calcite) is a mineral of extraordinary optical clarity. Its particular arrangement of atoms is responsible for a rhombic-shaped crystal that splits light rays to produce double images. As we have seen, calcite is also found in marble. As Iceland spar, calcite is crystallized into a different structural arrangement than it is in marble. It is this unique structural arrangement that gives Iceland spar its unique optical properties.

Obtain a piece of Iceland spar and a quartz crystal from a supply house (see the Appendix). Hold the Iceland spar in your hand under a light (or in sunlight) and notice the appearance of a colorful rainbow. Does a quartz crystal produce the same effect? Try a diamond (with permission of its owner). This rainbow effect is *dispersion* —the scattering of light by refraction forming a color spectrum. (A rainbow occurs in the sky when tiny water droplets disperse sunlight.) Certain minerals possess this special optical property. In Chapter 1 we saw how light is refracted (bent) when it passed from one medium (air) into another medium. Does Iceland spar refract (bend) light? Does the quartz crystal refract the light rays reflected from a sharpened pencil point if placed in front of it?

- Place an Iceland spar crystal with a large, flat surface over this sentence. What do you observe? Does the crystal magnify objects the same way as a magnifying glass? Does a quartz crystal give the same double-image effect or magnify objects?

SEVERING ATOMIC BONDS

Wearing safety goggles, use a hammer to break up an Iceland spar and a quartz crystal. Use the magnifying glass to observe what has occurred when each crystal is broken. When a mineral breaks, the atomic bonds are severed. If a mineral breaks along a smooth, even surface, it is said to have *cleavage.* If the break results in a mirror-like smoothness, the mineral has *perfect cleavage.* Does Iceland spar have perfect cleavage? Do the smaller pieces retain the same optical properties as before?

Use the magnifying lens to observe pieces of the quartz crystal. Quartz *fractures* (breaks unevenly) instead of cleaving. Study individual pieces of quartz with your lens. Note that each piece has edges that are rounded and curved.

If you grew large enough salt (halite) crystals (in the "Growing Macrocrystals" activity of this chapter), try the

103

"hammer cleavage test." Does halite exhibit fracture or cleavage?

READ MORE ABOUT IT

Bently, W., and W. Humphreys. *Snow Crystals.* New York: Dover, 1962.

Halfer, J. *How to Know the Rocks and Minerals.* Dubuque, Iowa: William C. Brown, 1970.

Lachapelle, E. *Field Guide to Snow Crystals.* University of Washington Press, 1969.

Mottana, A.; R. Crespi; and G. Liborio. *Simon and Schuster's Guide to Rocks & Minerals.* New York: Simon and Schuster, 1978.

Pough, F. *A Field Guide to Rocks and Minerals,* 4th ed. Boston: Houghton Mifflin, 1976.

Stangl, J. *Crystals and Crystal Gardens You Can Grow.* New York: Watts, 1990.

7

EXAMINING DOCUMENTS

A document is any writing that conveys information. Criminologists routinely use the magnifying glass to analyze writing materials: papers, ink, pencil, and type. Most of this analysis work emphasizes comparison of materials and writing with known standards to determine whether a particular document is authentic or forged; or perhaps it contains a hidden message? Let's look closer!

LOOKING AT PAPER

Paper is made up of plant (cellulose) fibers, sometimes with various additives to control the physical characteristics of the finished product. The choice of fiber is important. High-quality papers use cotton, linen, or hemp fibers —all of which provide for strength, stiffness, and durability. In the past, most book paper was made from straw or grass; today it is made from softwoods such as spruce or pine or the hardwood eucalyptus.

Paper begins with *pulp,* a material prepared chiefly from plant materials by chemical or mechanical means.

In *chemical pulping,* wood (or other plant material) is cut into small (5/8 × 1/8 inch) shavings or chips, which are cooked at high temperature, then treated with chemicals (sodium or calcium sulfite and sulfurous acid). This process separates individual, whole, cellulose fibers and removes lignin. *(Lignin* is a complex organic substance bound with cellulose to form woody tissue. If lignin is not removed from cellulose fiber, light will slowly turn it yellow-brown.) The extracted fibers are washed and bleached before being processed, usually through a paper machine. (Chemical pulps are essentially separated, whole cellulose fibers.) This book is printed on paper made from chemical pulp.

In *mechanical pulping,* bundles of wood fibers are mechanically torn from a debarked log (instead of being cut during chipping), producing a mixture of whole and broken fibers, lignin, and various other woody resins. Newsprint is made from mechanical pulp.

FIBER

An average wood fiber is 0.1378 inches (3.5 mm) long and around 1/4000-inch (0.06 mm) in diameter. It generally resembles a toothpick in shape. Use the magnifying glass to determine whether a paper sample is chemically or mechanically pulp-processed. Carefully tear or separate a small piece from a paper sample and observe individual fibers along the torn edge against a black background (use black construction paper). If available, use fine watchmaker tweezers to remove individual fibers for observation. Are individual fibers of equal length in each sample? Do your findings agree with your original deduction?

PAPER GRAIN

Observe different types of paper on a flat surface using direct lighting. Look carefully to identify individual fibers. *Paper grain* is the direction in which most fibers lie when a paper is machine-processed. Which way is the grain running on the paper of this page? Is there a difference in paper grain between book paper and sheets of writing or copier paper?

FIBER AND CLAY

Look carefully at paper samples with your magnifying glass. In writing paper (or the paper this book is printed on) you will see micro-hills and valleys that conform to the knitting of individual fibers that compose the paper. Compare these features to paper from a weekly magazine or calendar with color photographs. Do you notice a difference? Such "glossy" paper stock has China clay (a fine, white clay, kaolin) incorporated into it. Examine individual fibers from such glossy paper stocks to determine if the clay is mixed into the pulp or applied as a coating.

"OLD YELLOW"

Poor quality paper yellows. Study this degradation process by placing various paper samples in direct sunlight for an extended time period (four–eight weeks). Do you notice yellowing? What do you observe under the magnifying glass? Can you find individual fibers coated with lignin that are yellow or brownish yellow? What kind of light causes the fastest yellowing—the ultraviolet rays in sunlight or its other spectral colors? (Try using colored cellophane.) Will incandescent or fluorescent lighting cause newsprint yellowing?

Experiment with methods to slow down the yellowing process. Try dipping some small pieces of newsprint in a glass of water containing 2 teaspoons of borax (sodium borate) and allowing them to dry. Do these treated pieces yellow as quickly? What happens when you use vinegar instead of borax?

• You can artificially "age" paper by placing it on a cookie sheet in an oven and heating it at 150°F for one to two hours. (**This activity should be done only under adult supervision.**) Do you observe similar results as compared with your other experiments? How does the application of heat (instead of light) change individual paper fibers? Could you use the magnifying glass to discover that a document has been artificially aged?

RECYCLED PAPER

More than ever, recycled paper is being used as a printing medium. Use the magnifying glass to compare recycled paper with "first-use" paper of the same type, e.g., letter or card stock. Are color images printed on recycled paper as clear and crisp as those printed on glossy or nonglossy first-use paper?

PAPER CURRENCY

U.S. banknote paper is manufactured under heavy security at Dalton, Massachusetts, by a secret process. Banknote paper is unique, differing from high-quality fiber, or bond, paper in that it is 50 percent linen and 50 percent cotton. High-grade bond paper is made of cotton. Banknote paper must be able to withstand 2,000 double folds without breaking. If you have the patience, try folding a one-dollar banknote along with high-quality bond paper to see which "fails" first. Examine your work with your hand lens. Try to identify linen and cotton fibers.

During the papermaking process tiny colored silk threads are added to the pulp to make the paper more difficult for counterfeiters to copy. How many different-colored threads can you find with the magnifying glass? Are color fibers used by other countries in their paper currency?

WATERMARKS

Hold individual samples of high-quality bond paper, available at stationery stores, up to a light. Do you observe any markings? These translucent markings are the *watermark* and are used by paper manufacturers to identify their product, or a product for a particular customer. The watermark can help establish the approximate age of a document and possibly the location of the store selling the paper.

When you find a watermark, use the magnifying glass to examine both sides of the paper. Is there an equal indentation on each side? Is there a difference in spacing among fibers of the watermark as compared with those of the rest of the paper sample? Find out if watermarks are used for security purposes. Can you find watermarks in paper currency?

MAKING YOUR OWN PAPER

What You Need
>   white scrap paper (uninked white or colored)
>   plant or vegetable scraps
>   staples, tacks
>   2 wooden frames, suggested dimensions 4 × 5
>   inches (100 × 130 cm)
>   nylon fly screening
>   kitchen cloths (at least 4)
>   plastic washtub
>   mayonnaise jar
>   blender
>   sponge
>   iron

What To Do
Use the instructions listed below and Figure 24 as a guide to make your own paper.

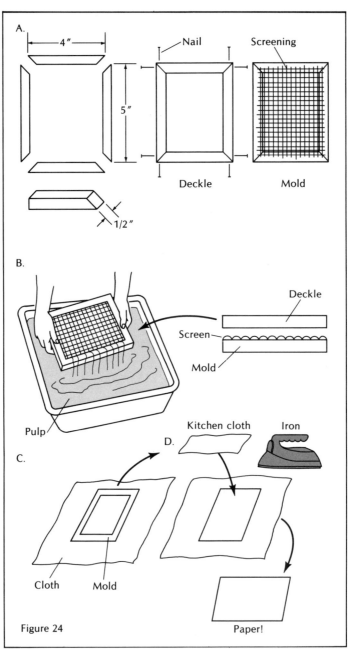

A.

4"

5"

1/2"

Nail

Deckle

Screening

Mold

B.

Deckle

Screen

Mold

Pulp

C.

Kitchen cloth

Iron

D.

Cloth

Mold

Paper!

Figure 24

Making your own paper

1. First you must make a paper mold. Tack or staple the fly screening tightly to the frame.
2. Take white, uninked scrap paper, and remove any plastic or staples. Tear it into very small pieces (about 1/2 inch, or 2 cm, square) and allow to soak in a mayonnaise jar filled with very hot water for half an hour. *(Note:* Allow the water to cool to a temperature that allows for comfortable hand immersion before going on to step 3.)
3. Take one handful of the soaked paper and place it in a blender about half full of warm water. Blend at a moderate speed until you can no longer see pieces of paper. (If you have problems, remove some paper.) To this mixture (pulp) add small pieces of finely chopped dry grass and blend again. If you wish, you can color the pulp with vegetable dyes.
4. Pour the pulp into a washtub half full of warm water. Increasing or decreasing the amount of pulp will affect the thickness of your paper.
5. Place the uncovered wood frame (deckle) on top of your screen. With both hands, dip the mold into the tub and scoop up some pulp. Gently shake the mold back and forth to get an even layer of fibers on the screen. When the water has drained completely through the mold, carefully remove the deckle, leaving the just-formed sheet on the screen.
6. To remove the paper from the screen, lay a clean kitchen cloth on a flat surface such as a table, then take the screen and lay it face down on the cloth. Soak up any residual moisture from the *back* of the screen with a sponge. Very *gently* lift the screen—the paper should remain on the cloth.
7. To dry the paper, cover it with another kitchen cloth and iron it at a medium-dry setting. Once dry, pull gently on either side of the cloth to stretch

it. This will help loosen the paper from the cloth. Gently peel the paper off.

Do not pour the remainder of the pulp down the sink—it might block the drain! The strained pulp can be thrown out or kept in a plastic bag in the freezer for next time.

- Use the magnifying glass to examine the finished paper sheet. Do you observe any paper grain? How does the fiber knitting in your handmade paper compare with that in commercial paper? Can you identify individual grass fibers? What would you have to do to the grass before it can be used to make paper?
- Try making paper from cloth alone. Use the procedure outlined above. This time separate individual threads from small (5 mm) pieces of cloth. Do your results compare favorably with paper advertised as having "100 percent rag" content when both are examined under a magnifying glass?
- Try recycling newsprint or glossy magazine paper by the same process as outlined above. How does paper made from these sources compare with uncoated white paper? Read about how recycling centers process (de-ink) newsprint before it is used again to make recycled newsprint.

INK AND PENCIL

Inks are composed of a number of substances (colored pigments) in a liquid solvent. Ball-point inks are thick solutions of organic dyes in an organic solvent like glycol or glycol ether. Other modern writing inks use a mixture of dye pigments in water. Printing inks use either a single pigment or a combination of pigments in a solvent such as kerosene. Inks used at the time of the first printing press used carbon black mixed with soap and linseed oil. Black writing inks of that period (1400s and later) contained pigments such as carbon black or iron gallonate. You can tell whether carbon black or iron gallonate ink was used on an old document. Inks that contain iron gallonate change from black to brown as the document ages. Inks containing carbon black do not change color with age.

- Use the magnifying glass to observe how various inks "lay" on various papers. Is the ink soaked in by the paper fibers or does it seem to rest above the paper's surface? When applying the same ink, is there a difference in image quality between coated and uncoated paper; between a porous paper such as newsprint and fine stationery paper?
- Make ink markings on various paper types to construct "standards." Label each marking as to paper type and ink applied. Observe how inks from ball-point pens lie differently than those from marker or fountain pens. Have a friend use the same pens used in making the "standards" to make "unknowns" for you to identify. They might even throw in a "ringer"—an ink that you don't know about—to test your identification ability!
- Try "aging" paper and writing samples. Make markings using various inks on white bond paper. With your parent's permission, place these marked sheets on a cookie tray and heat them in an oven at 150°F (no higher!) for one to two hours. Examine the paper with the magnifying glass to see what effects the heat had on both paper and ink. Can you convince your friends that you possess rare documents; if you did not know better, could those documents fool you?
- If possible, examine old documents and try to determine whether the ink used contained carbon black or iron gallonate.

The "lead" in a pencil is no longer lead! It is a mixture of graphite and clay with kaolin (aluminum silicate) as the binder. The more carbon (graphite) there is in the "lead," the softer the pencil mark; the more clay there is, the harder the pencil mark.

You can use the magnifying glass to determine if a pencil mark is made with a #1, #2, or #3 pencil. You can also determine the sequence in which pencil marks were made by examining the flow of graphite on paper. To find out, read on!

- Sharpen three pencils—#1, #2, and #3—to equal sharpness and place a 2-cm mark from each pencil type on a sheet of paper so that they are relatively close together. Label each mark. Use the magnifying glass to examine the marks. Have a friend select one of the three pencils and make a mark on another sheet of paper. Use your analytical skills to determine which pencil type made the mark.
- Have a friend place a line on a piece of paper using a #2 pencil while you are out of the room. Have him or her draw another line through it. When you return, arbitrarily label the two lines with letters. Can you determine which one was written first? The second mark should have "dragged" some of the other mark with it.
- Another method to determine the order of marks is to use Silly Putty. Flatten some Silly Putty so that one surface is very smooth. Place that surface over the crossed pencil marks from above. Press down to get a good imprint. Carefully observe this replica with the magnifying glass from all angles to see the different impressions from the marks so that you can determine which was made last. Remember that the marks will be reversed!

TYPEWRITTEN DOCUMENTS

Type comes in all sorts of sizes, shapes, styles, and colors. Usually an individual typewriter can be identified by its own unique type. Try to get a complete set of type characters (one set per page) from a number of different typewriters. Use the magnifying glass to analyze the following characteristics of type for each machine: type size, spacing between letters and irregularities (broken type). Record your observations in your notebook.

- Does each typewriter appear to be unique? Could you identify a given typewriter by its type if a friend tested your abilities?
- Can you use this process to analyze type from a dot-matrix printer or laser printer? Why?

HANDWRITING ANALYSIS

A person's handwriting is somewhat like his fingerprints: it contains certain individual characteristics. If a person tries to change his handwriting, he will not succeed! Handwriting is the product of two processes: physical movement, and the subconscious activity of the mind.

When someone tries to forge another person's name by tracing, this act can be detected by the magnifying glass. Have a friend write his or her name on a small sheet of paper and label it O for original. Have the same person write the name again and label it A. Another person should trace the signature onto another sheet of paper; label it B. A third person should try to forge the original by just looking at it and writing it down. Label this C.

Use the magnifying glass to observe the following for each writing samples B and C: the line quality—is there evidence of retracing or retouching; are the slant and cursive (the writing flow) as well as the proportion of the letters similar to or different from either the original or the duplicate, A? Can you tell why B and C are forgeries?

EXAMINING MICROPRINT

Microprint is a tool used by printers of checks and other negotiable instruments to prevent forgery. Each character in a word is a microscopic high-resolution graphic image. This security feature cannot be duplicated using copiers or desktop scanners. To the eye (as well as copiers and scanners), a line of microprint characters appears as a solid line. This is the size of the microprint:

(Courtesy of Standard Register)

READ MORE ABOUT IT

Heller, J. *Papermaking.* New York: Watson-Guptill, 1978.

Santoy, C. *The ABC's of Handwriting Analysis,* New York: Paragon-House, 1989.

Smith, E. S. *Paper.* New York: Walker, 1984.

Studley, V. *The Art and Craft of Homemade Paper.* New York: Van Nostrand Reinhold, 1977.

8
FABRICS AND FIBERS

Fabrics (or cloths) are made by interweaving one set of threads (yarns) with another on a machine called a loom. A thread or yarn is a continuous or plied strand composed of filaments, or *fibers*.

Until this century, natural fibers such as linen (from the flax plant), wool, cotton, and silk (from cocoons of silkworms) were the basis of woven fabrics. Today, many fabrics combine polyester (a synthetic polymer fiber called Dacron) with natural fibers such as wool or cotton. Some synthetic fibers, like nylon and rayon, are used exclusively in fabrics as well as other articles such as toothbrush filaments and artificial turf.

INVESTIGATING WEAVINGS

The three most common ways of weaving patterns in fabrics are known as plain, twill, or satin.

Use your magnifying glass to examine a handkerchief. In your notebook draw the pattern you observe. Notice that this regular pattern—or *plain weave*—con-

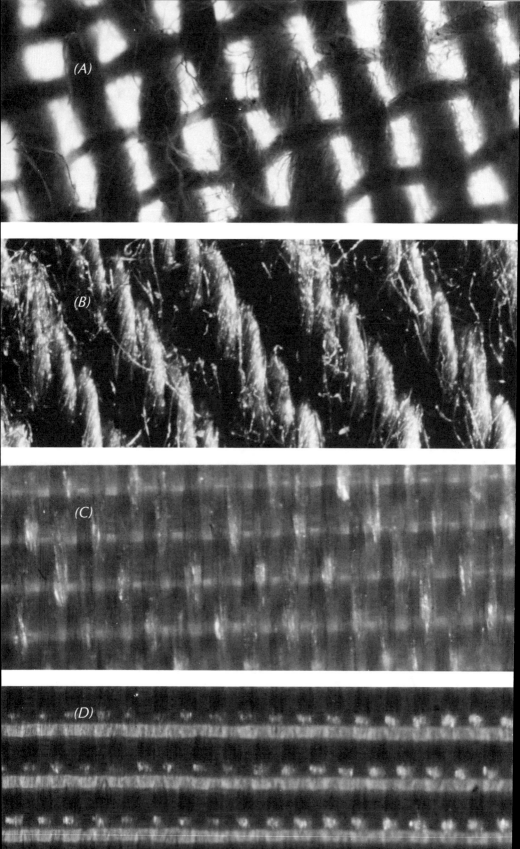

sists of cross threads (the weft or woof) going first over and then under lengthwise threads (the warp). Many cottons and linens are woven in this plain weave pattern. Examine this weave pattern in table linen, dress shirts, woolen tweed, and gingham fabric with your magnifying glass.

The *twill weave* is made by crossing lengthwise threads with crosswise threads in an irregular way so that the finished fabric has rows of diagonally raised lines. This weave pattern gives longer wear to the fabric. Use your lens to examine business suits, which are usually made from this type of cloth.

The *satin weave* is really a broken twill pattern. In the satin weave, twill lines do not show. Instead of progressing *one* warp thread at the beginning of each new crossing of the weft threads, it progresses in *two*. This fabric gets its smooth appearance on one side because the many warp threads hide the weft threads. Use your lens to examine silk fabrics, which are almost exclusively made using this weave.

TESTING FABRICS

Fabric testing, often under rigorous conditions, is routinely done in the textile industry. Here are some tests you can perform at home:

• *Quality.* Generally the number of weft threads running in a square inch of fabric is a good indicator of

Weave patterns.

(A) Plain weave.
(B) Twill weave. (C) Satin weave (fine weave). (D) Satin weave (bulky weave).

quality and long wearability. Try counting wefts of different fabrics. Does the cloth material in a handkerchief have more or fewer wefts per inch than your dress shirt?

- *Creasing.* A fabric's fiber content may be seen by creasing. Linen creases very easily, and remains creased longer than cotton. Pure wool or silk resists creasing and tends to spring back into shape. (This test cannot be used if the fabric has been chemically treated, or "sized.")
- *Strength.* Use your magnifying glass to examine fabric weave patterns for unevenness. These thinner spots will wear faster. Examine articles of clothing already having definite areas of wear. Describe any differences you observe.

EXAMINING FIBERS

Another method of testing fabrics is to unravel and examine individual threads or yarns. Use tweezers under the magnifying glass to remove warp and weft threads; separate each type. Hold individual threads with tweezers while examining them under your glass. Try unraveling threads to expose individual fibers. (**Always ask permission before examining fabrics from garments.**)

Here are some things to look for:

- *Cotton.* The threads look dull and limp unless they have been mercerized. *Mercerizing,* treating with sodium hydroxide, gives cloth added luster and strength and allows for better dye retention.
- *Linen.* The threads are stiffer and more lustrous than unmercerized cotton threads. Linen also absorbs moisture readily. A common test for linen is to moisten a finger and place it under the fabric while

viewing with a magnifying glass. How does sizing affect the water absorbency of linen?

- *Wool.* The threads are springy and lustrous. Individual fibers are crinkly.
- *Silk.* The threads are smooth and highly lustrous. Individual fibers are very thin and fine.
- *Synthetics.* Use your magnifying glass to examine women's hose (stocking) threads. These threads are usually made of nylon. How do they compare to the natural fibers you examined earlier? Look at other synthetic fabrics such as rayon, polyester, etc. (Use the garment label to tell you what each fabric is made of.)

FABRIC COLOR

For thousands of years man has improved fabrics and other materials by dyeing them different colors. *Natural dyes* are made from plants and animals; *synthetic dyes* are chemically compounded and are now mostly used in the textile industry. Dyes can be applied to the raw material before it is spun into yarn, to the yarn after it has been spun, or to the fabric after it has been woven.

Use the magnifying glass to examine color patterns or areas in fabrics. Can you determine if the colors were already present when the fabric was woven, or applied after weaving?

Remove individual colored threads and examine them under your glass. Are all the individual fibers uniformly dyed?

DYE YOUR OWN

Take small pieces of various fabrics and dye them following the directions given with commercial fabric dye products, which are available in most supermarkets. **Do this**

under adult supervision. Use the magnifying glass to examine individual threads and fibers. Do different fibers react differently to dyeing?

Dye fabric pieces with the natural dye anthocyanin. **Under adult supervision,** boil some beets or red cabbage for five minutes to extract this natural plant pigment. Strain the red-colored water of any plant material and allow to cool. Pour the colored water into three glass jars. To one jar slowly add white vinegar. What happens? To another jar slowly add baking soda. Is the color change different? Do you suppose you can reverse these color changes by adding additional baking soda or vinegar?

Immerse fabric pieces in each of these colored solutions for at least twenty-four hours. Remove and let dry. Use your lens to examine individual threads and fibers. Are these natural plant pigment colors "fast" dyes—i.e., do they remain following washing with water?

PERMANENT PRESS AND OTHER MATTERS

Permanent press garments are treated with a chemical (usually formaldehyde) so that they do not require ironing following washing. Formaldehyde "fixes," or crosslinks, natural fiber protein so that treated fabrics will retain a crease. Examine threads and fibers from such garments (with your parent's permission) and compare them with untreated examples of the same fiber type.

Purchase a commercial waterproofing agent (at supermarkets) and apply it, according to directions, to a small area of a handkerchief. Mark this area with an indelible marker. Use a larger magnifying glass (2-inch diameter) to examine what occurs when you spray the entire handkerchief using a water-spray bottle. Use the magnifying glass to examine individual threads. Do you notice any difference between treated and untreated areas? Examine the handkerchief after it has been washed. Notice any changes? Wash and examine again. Is this waterproofing treatment permanent or temporary?

Examine fabrics that are advertised as waterproof. Is it the weave or a chemical treatment of individual threads that makes the fabric repel water?

STAIN REMOVERS

There are numerous advertised products that remove "the toughest stains." See if you can duplicate these advertising claims by applying similar stain agents to a stained handkerchief. Examine the fabric before and after treatment with a stain-removal product. **(Follow the manufacturer's directions.)** Use the magnifying glass to learn how stains are trapped by both the cloth weave and individual fibers. Do certain fiber types hold the stain and resist cleaning? Do these fabric cleaners work? Which cleaner works the best? Also try staining individual fibers followed by immersion in cleaning solutions to observe how they are affected.

Some stain-removal products contain special protein compounds (enzymes). Use your magnifying glass to examine treated fabrics to see how they work. Are individual threads of natural fiber fabrics weakened if a concentrated amount of this type stain remover is allowed to remain in direct contact for a long time period?

Compare commercial stain removers to a natural material, yogurt. Yogurt contains natural enzymes (just like the ones advertised in some commercial products) that are capable of degrading a stain. Apply yogurt by rubbing it into the stained area; remove the excess with a butter knife. Repeat until the stain is removed. Does this stain-removal process weaken fabric fibers?

READ MORE ABOUT IT

Goodwin, J. *A Dyer's Manual.* London: Pelham, 1982.

McRae, B. *The Fabric and Fiber Sourcebook.* Newton, Conn.: Taunton Press, 1989.

9
SURFACES

Surfaces are the upper boundaries of any material body. From the perspective of standing on the moon, the earth looks smooth; but we all know it isn't! A concrete floor certainly appears smooth to us; but what if we could see it from an ant's perspective!

The magnifying glass will provide you with this unique perspective, the ability to view surfaces to see just what these outer boundaries really are!

IS "SMOOTH" REALLY SMOOTH?

To the eye, the close inspection of the flat surface of a wooden board purchased at a lumberyard appears smooth. Gently run your hand across its flat surface. As you move your finger, the touch receptors in your fingertip relay impulses to your brain to contradict your visual observation. A closer look through your magnifying glass demonstrates that the board is indeed not smooth!

Obtain sandpaper from a hardware store; try to get

124

at least three types: coarse, medium, and fine grit. Sandpaper is a cloth or paper sheet coated with an *abrasive,* a substance used to smooth, grind, or polish materials. Examine each grit type with your lens. Can you relate the size and number of individual grains with the sandpaper type? Are grit size and grain number consistent? With your glass, can you identify the type of grit material used in making sandpaper? (*Hint:* Look at beach sand.)

Try looking at other abrasive materials with your lens: whetstones, grinding and polishing wheels, emery cloth. Why is each type of abrasive usually specific to a particular need—would you use sandpaper to smooth or polish metal?

Use some sandpaper to smooth a board's surface. Start with the coarsest-grit sandpaper; after sanding examine your work with your lens. Work through the remaining grit sizes. Is your board smooth? By what measure of perspective: eye, finger, or lens?

Here are some other "smooth" activities you can do with a magnifying glass:

- Apply a paint primer to an area of your sanded board; allow to dry. Now apply a latex or oil-base paint over both primed and unprimed (but sanded) areas. Allow to dry. Does priming make for a smoother surface?
- Examine a piece of sheet glass placed over a piece of black construction paper. Is it smoother than a new sheet of acrylic plastic examined in the same way? Smoother than the top of your kitchen counter?
- At an auto-parts store, obtain some windshield rain repellent (a mixture of the water repellent silicone, and alcohol). Apply this material to half of the sheet glass (or a mirror) following manufacturer's directions. Next use a mist-spray bottle to *lightly* spray water over the entire surface of the glass

sheet, lying flat over a piece of black construction paper. Use a large-diameter magnifying glass to carefully examine water droplets. Do you notice a difference? Now raise the glass sheet and observe what happens. Would you recommend this product?

- Use your lens to examine the way leaves, stems, and flower petals both retain and shed water droplets following misting (with a spray bottle), or after a gentle rain.
- Find out how the application of polyurethane coats makes furniture surfaces smooth.
- Does the application of spray polish on fine furniture surfaces create "waxy buildup"? **(Be sure to get adult permission before you start!)**
- Under adult supervision, try using pumice (volcanic stone), available at hardware stores, to polish nail shanks. Examine the nail with your lens before and after polishing.
- Are pinpoints sharp or dull?

IS "ROUGH" REALLY ROUGH?

By eye, the face of a file certainly appears rough. By touch as well! Examine the face of a file with your magnifying glass. Notice the cutting ridges. Is each ridge surface smooth or rough?

MOVING MOLECULES: INVESTIGATING SURFACE TENSION

Surface tension is a force that makes the surface of a liquid act as an elastic film. Surface tension is caused by cohesive forces that attract molecules of a liquid to each other. Surface molecules are attracted mainly downward, because there are no molecules of liquid above them.

Cohesion makes the surface molecules resist forces that change their position.

Obtain a small sewing needle and grasp it with tweezers. Now, very carefully lay it on the surface of a bowl of water set over a piece of black construction paper. The needle floats! Shine a light (from the side) on the water's surface. Use the magnifying lens to observe the behavior of the water film around the needle. While peering through your glass, use another needle to gently move the floating needle about. What happens to the water film? Now, while still looking through your lens, add a couple of drops of liquid dishwashing detergent to the bowl of water. What happens?

Try another experiment. Add a number of drops of salad oil to a shallow dish of water placed on black construction paper. As before, observe the surface with your magnifying glass. As you are examining the surface under the lens, add a drop of liquid detergent. What happens? Did the soap reduce the surface tension at the point where it was dropped onto the water's surface? Why were the oil droplets pulled away?

FOSSAE AND TRABECULAE—BONES

Bone is a remarkable material. It is eight times stronger than concrete, requiring over 12,000 pounds per square inch to crush it. To achieve this durability, bones are made of a composite material: calcium phosphate crystals, which withstand the forces of compression, wrapped around ropelike collagen protein fibers, which handle the forces of tension.

Let chicken leg bones dry out for a couple of days following your dinner. Use a coping saw to saw off, at an angle, this bone as illustrated (cut line 1) in Figure 25. Use your magnifying glass to study this end cut. Note a pat-

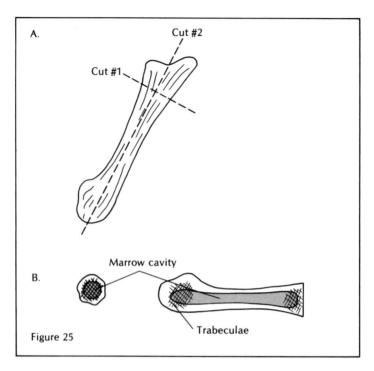

A.

Cut #2

Cut #1

Marrow cavity

B.

Trabeculae

Figure 25

Inside a chicken bone

terned meshwork of tiny spicules, or trabeculae, that are present. Trabeculae reinforce bone ends against stress. They function in much the same manner as the tension cables installed on suspension bridges like the Brooklyn Bridge.

Now use the hacksaw to saw your leg bone in half, lengthwise (see cut line 2 in Figure 25). Is it solid? Notice that the ends of long bones (such as this chicken femur) are not solid but hollow. This cavity contains *bone marrow,* which replenishes blood cells. Use your magnifying glass to study the interior marrow cavity. Do you notice trabeculae?

Use the magnifying glass to study the surface of the leg bone. Notice the numerous tiny depressions (fossae), holes (foramen), and bumps (tuberosities). Each has a special significance in aiding the attachment of muscle and ligament, as well as in movement. Note that the bone's surface is not smooth but delicately pitted.

Use the reference books at the end of this chapter to learn how to prepare animal skeletons. You can also visit a local museum to observe and study prepared skeletons there. Examine the equivalent bone from different animals and compare how it is similar or dissimilar. Does a bone's form follow its function?

THE CUTTING EDGE

Use your lens to examine the edge of various cutting surfaces: safety razor, scissors, penknives. Are the cutting surfaces of each of these articles always uniform? Research what types of metals best hold a sharp edge. **With your parent's permission,** try sharpening a knifeblade using a whetstone. How fine an edge can you get?

Scissors have a less sharp edge than a razor or even a penknife. How can they cut paper?

COSMETIC EFFECT

Cosmetics make women's faces more attractive. Certain forms (eye shadow, some lipsticks) contain minute mineral (mica) particles that provide an iridescent effect. Apply some to a glass plate over black construction paper and observe them with your lens, using side lighting.

NATURAL SURFACES

Like the naturalists of this and the previous centuries, use your magnifying glass to examine natural surfaces. Draw what you observe in your notebook. Here are some places to start:

- How are certain seeds (dandelion, goldenrod, maple, and pine) carried by the wind? How are some seeds (cocklebur, beggar's tick, and tick trefoil) carried by you?
- How is a spider web constructed? (Spray the spider's net with white spray paint. Catch the net against a piece of black felt previously sprayed with clear lacquer.)
- Spore cases (sori) of ferns (undersides of leaf or frond.)
- Black spot (a fungus) on maple leaves in late summer and early fall and/or sooty spot (another fungus) on clover leaves.
- How do seashells get their color?
- How do wooden surfaces weather?
- Early American colonists used the horsetail *(Equisetum)* as a natural scouring pad. Examine shoots of these primitive plants to find out why.
- Do natural sponges have more holes than artificial, commercial ones?

READ MORE ABOUT IT

Gale, L. *The Indoor Naturalist: Observing the World of Nature Inside Your Home.* New York: Phalarope Books (Prentice Hall), 1986.

How to Prepare Skeletons. Rochester, N.Y.: Ward's Natural Science Establishment, 1962.

Stokes, D. *A Guide to Observing Insect Lives.* Boston: Little, Brown, 1983.

10
A CLOSER LOOK AT YOURSELF!

The magnifying glass can reveal a lot about who you are, and why you are unique!

NAILS

Nails are broad, slightly curved plates on the top surface of mammalian digits (fingers and toes). Without them, it would be very difficult for you to pick up objects. Mimic the absence of fingernails by wearing oversize gloves with padding in the fingertips. Use these gloves to pick up small objects—a penny, a wood screw, a piece of paper. After this exercise you should have a greater appreciation for your nails! Take a moment to look at photographs of different mammals having nails. Imagine them without nails. Do you think they could survive?

Trim a small piece of a fingernail using sharp scissors or a nail clipper. Cut this piece into several smaller pieces. Examine it with your magnifying glass. Your nail is made up of three plates or layers. With a sewing needle, probe the upper, middle, and lower surfaces and notice that the middle portion (*middle nail plate*) is softer than the other two layers. The top surface (called the *upper nail plate*) is much harder and tougher than the other layers because of its high calcium content. The bottom, or *lower nail plate,* provides a foundation or support layer covering the surface of the finger. With a fine forceps, try scraping or tearing apart the piece of nail and notice where it frays or flakes. What do you think happens when you injure a nail?

INVESTIGATING YOUR BODY'S LARGEST ORGAN—SKIN

Your skin makes up about 16 percent of your body weight. As your body's largest organ it functions to protect you from injury, helps retain moisture and fluids, receives stimuli from the environment, excretes various substances, and helps keep you warm.

Use the magnifying glass to examine the surface of your forearm. You can see that the surface of your skin certainly is not smooth! Observe the many delicate grooves, or *flexure lines,* that create patterns on your skin.

Notice the many body hairs present. Body hair is present in all mammals during some part of their life. Look closely at individual body hairs on your body. Use the magnifying lens to examine the palms of your hands and the soles of your feet. Notice any body hair? Under the magnifying glass, do you notice any difference between your body hair and the hair on your head?

STUDYING CYTOMORPHOSIS—SKIN FLAKING

Deep within your skin are cells that are constantly growing. As new cells are created, older ones are continually pushed upward—a process that takes from fifteen to thirty days on average.

Use your magnifying glass to search for dead or flaking skin areas or particles. You can speed up the process by making a *light* mark on your skin using your fingernail. Look closely with your lens to see tiny areas of dead skin that dislodged.

THE UNIQUENESS OF SELF—FINGERPRINTS

The study of dermatoglyphics is also known as the science of fingerprinting. It is an old science, known by the ancient Chinese, who recorded it in drawings on official seals.

Fingerprints and toeprints are caused by the development of tiny ridges in the outer skin. The arrangement and number of these ridges are unique to each individual. Because of this, forensic scientists (those who study crime) use fingerprints to assist in identifying suspects in criminal cases.

Fingerprint patterns may be grouped into three large categories: the *arch,* the *whorl,* and the *loop.* In the arch pattern the ridges extend across the bulb of the finger and rise slightly at the center. Ridges in the whorl pattern create a spiral or circle on the finger. The loop pattern consists of one or more ridges curving into a hairpin turn. When experts compare two fingerprints, they usually require ten to twelve points of similarity between the two to establish that the prints are identical, that is, from the same individual.

There are so many specifics of fingerprint classification and cross-indexing that it is impractical to deal with those subjects in this book. Should you wish to learn more about this fascinating science, read *The Science of Fingerprints* (publication 1988-241-721/94805), written by the Federal Bureau of Investigation (FBI). You can get a copy of your own by writing to the Superintendent of Documents, U.S. Government Printing Office, Washington, DC 20402.

For beginning fingerprint analysis you will need a blank ink pad, a number of white, unlined 5 × 7 inch index cards, and a magnifying glass.

To make a reference set of fingerprints, label the top of an index card as follows:

Name _____   Date _____

Left Hand

Thumb	Index	Middle	Ring	Little

Right Hand

Thumb	Index	Middle	Ring	Little

Use a single card for each set of an individual's fingerprints. These types of prints are called *direct prints* by forensic investigators and are similar to those retained by the FBI. (Currently, the FBI has over 174 million prints on file in its Fingerprint Identification Division. Automatic fingerprint reading devices (computer scanners) can scan a fingerprint and match one to its file in seconds!)

To take a fingerprint impression carefully roll each fingertip across the surface of the ink pad; then roll it in the center of the appropriate finger-image area on the index card. Make sure that you obtain a clean set of impressions; otherwise repeat until you do. Wash fingers with soap and water before proceeding.

Use your magnifying glass to examine each fingerprint.

- Does each of your fingerprints have the same pattern?
- What differences are there in the pattern of your left and right index and thumb fingerprints?
- Try comparing your fingerprints to those of other members of your family; to friends'. Do you notice any similarities, differences?

Here are some other activities you can try using fingerprint analysis:

LIFTING LATENT PRINTS

Latent, or "hidden," prints are impressions caused by perspiration on the ridges of the skin making the print. Perspiration contains water, salt, amino acids, and perhaps oil, dirt, grease, or blood. The method used for obtaining latent prints depends upon the type of surface to be examined, the manner in which the prints were left, and the quantity of material left behind.

Do this activity with a couple of your friends. Take a reference set of direct prints from each individual as you did with your own.

Wash a glass and rinse it thoroughly with water. Wipe it dry, making sure no fingerprints appear on the glass surface. Hold the glass with a cloth or paper towel and have one of your friends (you should not know which friend it is) place his or her *right* thumbprint on the glass surface.

Use powdered graphite as your dusting powder. Place a small amount in a small cup. (***Careful!* Graphite is very slippery! Protect working surfaces with newspaper to collect any excess graphite. Do not inhale any graphite dust!**) Use a clean brush for dusting. Make sure the bristles are separated from each other. Dip the brush into the cup containing powdered graphite and lightly dust the area suspected to contain the print. After the print is developed, remove excess graphite powder by gently brushing it away. Be careful not to destroy the print with too hard a brush stroke! If you have a camera available, photograph the developed print.

To lift the print from the glass to the index card, unroll about 5 or 6 inches of clear sticky tape (not frosted) and place the end to the right of the developed print. Allow the rest of the tape to cover the print. Carefully smooth the tape over the developed print to force out all air bubbles.

Remove the print by pulling up on the free end of the tape and place it (print side down) on a clean white index card. Cut off the edges of the tape.

Observe the print with your magnifying glass. Compare this print to your file of "suspects." Can you get a match?

Use white talcum powder in place of powdered graphite when dusting for fingerprints on dark surfaces. Also use a different brush for each type of dusting powder. Graphite can also be used to dust fingerprints on light surfaces.

IDENTIFYING A DIRECT PRINT

Occasionally a criminal will leave behind a print that is clearly visible. Such prints are made by colored matter on the fingers: soot, ink, dyes, etc.

Have one of your friends make some direct prints on pieces of paper. Compare this print to those in your files. Can you make an identification? Can you definitely rule other individuals out?

LET'S SEE THOSE TASTE BUDS!

Besides assisting in food passage and in providing for articulated speech, your tongue allows you to discriminate by taste. There are four primary taste sensations: sweet, salty, bitter, and acid.

Use a large-diameter magnifying glass to study the surface of a human tongue. (You can examine a willing participant or examine your own by observing your reflection in a mirror.) Consult an anatomy text to assist you in identifying the three major types of *papillae* (raised bumps), which are structures that contain *taste buds* (microscopic groups of cells that connect by nerves to the brain). Notice that vallate papillae are the largest, looking like tiny mushrooms; they reside in a V-shaped area near the base of the tongue. Filiform papillae are very small and numerous, especially along the sides of the tongue. Fungiform papillae are larger but less numerous than filiform papillae; they are scattered toward the middle.

Have a partner dip a cotton swab into solutions to help map where certain taste sensations are located. Try sugar water, lemon water, salt water, and vinegar water. Mix two teaspoons of any of these substances in a glass of water. **Carefully** (avoid touching the cotton swab too close to the back of the tongue so that a gag reflex results) place the tip of the cotton swab on identified papillary areas of the tongue. Be sure to use a magnifying glass to help locate filiform papillae. Draw a map that indicates where certain taste sensations are concentrated.

READ MORE ABOUT IT

Federal Bureau of Investigation. *The Science of Fingerprints* (publication 1988-241-721/94805). Washington, D.C.: U.S. Government Printing Office, 1988.

APPENDIX
WHERE TO OBTAIN
MATERIALS

Magnifying glasses can be purchased at local stationery, camera, or art-supply stores. Much of the material and equipment needed to do the activities in this book can be obtained around the house or from the local hardware store. However, some materials—like Iceland spar and other mineral specimens, exotic insect forms, and collecting equipment—will need to be purchased from supply companies. The following firms offer a complete line of science materials.

Most supply companies will supply materials if an order is accompanied by a check. One company (Edmund) will provide a catalog if you call; ask to borrow your teacher's catalogs for information about other companies.

American Biological Supply Company
1330 Dillon Heights Avenue
Baltimore, MD 21228
(301) 747-1797
(Equipment and apparatus only)

Carolina Biological Supply Company
2700 York Road
Burlington, NC 27215
(800) 547-1733

Connecticut Valley Biological Supply Company
82 Valley Road
South Hampton, MA 01073
(413) 527-4030

Edmund Scientific
101 E. Glouster Pike
Barrington, NJ 08008
(609) 573-6250
(Equipment and apparatus only; many types of magnifying glasses)

Science Kit and Boreal Laboratories
777 East Park Drive
Tonawanda, NY 14150
(800) 828-7777

Ward's Natural Science Establishment
5100 West Henrietta Road
P.O. Box 92912
Rochester, NY 14612-9012
(800) 962-2660

INDEX

China clay, 107
Chitin, 55
Chromatophores, 61
Circuli, 61, 63
Cleavage, 97, 103–4
Cockroaches, 57
Cohesion, 126–27
Coins, 102
Color photographs, 31–32
Computer bar codes, 38
Concave lenses, 12
Contour feathers, 68
Convex lenses, 12
Copper, 101
Cork, 76
Cosmetics, 130
Cotton, 117, 119, 120
Crystals, 92–96, 102–3
Ctenoid scales, 59, 61, 63
Currency:
    metal, 102
    paper, 37, 108
Cuticle, 67
Cutting edges, 129
Cyanobacteria, 44, 45
Cycloid scales, 59–61
Cytomorphosis, 133–34

Degradation, 51–52, 107–8
Diamonds, 103
Direct lighting, 19–21
Direct prints, 135–36, 137
Dispersion, 103
Documents, 105–16
    handwriting on, 115
    ink and pencil on, 112–14
    microprint on, 116
    paper in, 105–12
    typewritten, 115
Double-convex lenses, 12, 13
Double images, 102–3
Down feathers, 68
Dragonflies, 56, 57
Duckweed, 48
Dyes, 121–22

Earthworms, 67–68
Effervescent minerals, 98

Engravings, 35–38
Enzymes, 123
Euhedral crystals, 94
Exoskeletons, 54, 56
Eyes, compound, 54, 56

Fabrics, 117–23
    color of, 121–22
    fibers in, 117, 120–21
    permanent press, 122
    stain removers and, 123
    testing of, 119–20
    waterproofed, 122–23
    weavings in, 117–19
Facets, 56
Feathers, 68–71
Feet, of mollusks, 64
Feldspar, 96, 97
Fertilizers, 81–83
Fibers, 106, 117, 120–21
Film speed, 33
Fingernails, 70, 132–33
Fingerprints, 134–37
Fish, 59–61, 63–64, 71
Flies, 57
Flowers, 75, 86, 87, 126
Focal length, 13, 14
Folding pocket magnifiers, 16
Food spoilage, 50–51
Forgery, 115–16
Fossils, 71–73
Four-color process, 31
Fracture, 97, 103–4
Fresnel lenses, 16–19
Fungi, 44, 50–51, 75

Galls, 84–86
Gamion, 59
Ganoid scales, 59
Glass, 125–26
Granite, 96, 97
Grass, 48, 81–83
Gymnosperms, 86

Hair, 55, 56, 64–67, 133
Hair scales, 67
Halftone, 31
Halite, 94, 103–4

141

Stamen hairs, 76
Stand magnifiers, 16
Starch grains, 90
Stems, 75, 79–81, 126
Stereoscopic images, 38–43
Stomates, 76–79
Stones, green, 45
Subhedral crystals, 94
Supersaturated solutions, 94
Surfaces, 124–31
    rough, 126
    smooth, 124–26
Surface tension, 126–27
Synthetic fibers, 117, 121

Taste buds, 137–38
Terminal buds, 84
Thorax, 54, 56, 57
Three-dimensional imaging,
    38–43
Tobacco, 90
Trabeculae, 129

Twigs, winter, 79
Twill weave, 119
Typewritten documents, 115

Vascular bundles, 79–81
Virtual image, 12

Water, microlife in, 45
Water lenses, 11–12
Watermarks, 109
Waterproofing, 122–23
Water samples, 24
Weavings, 117–19
Windshield rain repellent, 125–
    26
Wings, of insects, 54, 57
Winter twigs, 79
Wool, 117, 120, 121
Worms, 67–68

Yellowing, of paper, 107–8
Yogurt, 123